Solid Rocket Propellants
Science and Technology Challenges

Solid Rocket Propellants

Science and Technology Challenges

Haridwar Singh
Defense Research & Development Organisation, Pune, India
Email: sharidwar@gmail.com

Himanshu Shekhar
High Energy Materials Research Lab, Pune, India
Email: himanshudrdo@rediffmail.com

THE QUEEN'S AWARDS
FOR ENTERPRISE:
INTERNATIONAL TRADE
2013

Print ISBN: 978-1-78262-096-9

A catalogue record for this book is available from the British Library

Published by The Royal Society of Chemistry,
Thomas Graham House, Science Park, Milton Road,
Cambridge CB4 0WF, UK

Registered Charity Number 207890

Visit our website at www.rsc.org/books

Printed in the United Kingdom by CPI Group (UK) Ltd, Croydon, CR0 4YY, UK

Preface

We live in a space and missile age. For efficient and powerful propulsion of missiles and space vehicles, advanced and environmentally friendly propellants are needed that are insensitive to shock, friction and impact. Rocketry is an important branch of science and technology in the modern age.

This book attempts to describe and discuss various subjects on science and technology for solid rocket propellants; including modern topics. We sincerely believe that some of the most important and advanced topics like shelf life prediction, advanced solid propellants, quality control and reliability, safety during process and handling *etc.* are a unique feature of this book. These topics are of vital importance to scientists/engineers working in research and development establishments, production agencies, quality control groups, academia *etc.* This is perhaps the first book of its kind dealing with a large number of important topics of relevance today. We are confident that the utility of this book will be of high order not only in our country but elsewhere as well.

The book contains 12 chapters. Major highlights include a detailed description of solid rocket propellant processing technologies, insulation, liner and inhibition systems, ignition system, combustion mechanisms *etc.* A detailed description of topics such as thrust vector control, structural integrity, rocket motor casing materials, catalyzed and platonized propellants *etc.* are also included. To the best of our knowledge, this information is not available as a single source in earlier publications.

Solid Rocket Propellants: Science and Technology Challenges
By Haridwar Singh and Himanshu Shekhar
© Haridwar Singh and Himanshu Shekhar 2017
Published by the Royal Society of Chemistry, www.rsc.org

During the process of writing this book, we have taken valuable suggestions from a number of renowned scientists in India and abroad. We express our gratitude to all of them.

His Excellency, the Late Dr A. P. J. Abdul Kalam, Hon'ble Ex-President of India and founding father of Indian Missiles has been a great source of inspiration in preparing this manuscript. We express a deep sense of gratitude and respect to him for his valuable suggestions.

Dr Haridwar Singh
Dr Himanshu Shekhar

Dedicated to:
Dr A. P. J. Abdul Kalam
The father of the Indian Missile
Programme

Author Biographies

Dr Haridwar Singh
Emeritus Scientist (DRDO)
(Former Director & Outstanding Scientist,
HEMRL, Pune-21)

Born in 1944, Dr Haridwar Singh obtained his Masters Degree in Chemistry in 1966. He served as a Lecturer in Chemistry at the Government Post-Graduate College, Gyanpur, Varanasi from 1966–1970. Subsequently, he joined the High Energy Materials Research Laboratory as a Scientist. He earned a PhD degree of the University of Poona, for his thesis in the field of Solid Rocket Propellant Combustion. Dr Singh is a recognized Post-Graduate Guide of the Poona University and has so far supervised a total of thirty Doctoral and Post-Graduate theses. He has more than 150 publications, eight patents and two books on "Rocket Propellants" and "High Explosives" to his credit. He has participated and chaired technical sessions in a number of international/national seminars and workshops.

Dr Singh was a Visiting Scientist at the Max-Planck Institute, Gottingen, Germany; Pennsylvania State University; University of Arizona; Army Research Lab, Maryland and Sandia National Lab, Livermore in USA. He served as an expert member of a specialist group, constituted

Solid Rocket Propellants: Science and Technology Challenges
By Haridwar Singh and Himanshu Shekhar
© Haridwar Singh and Himanshu Shekhar 2017
Published by the Royal Society of Chemistry, www.rsc.org

under OPCW, Hague. Dr Singh was the President of the High Energy Materials Society of India (HEMSI), a Senior Member of the American Institute of Aeronautics and Astronautics (AIAA), Fellow, Maharashtra Academy of Sciences, Senate Member, Pune University, Member, Editorial Executive Committee, Defense Science Journal, Member, Research Board, American Biographical Institute. He is also the Honorary Member of the Russian Academy of Astronautics.

Dr Singh's major contributions include the establishment of advanced indigenous technologies for cast double base (CDB) propellants, high energy nitramine based propellants, Composite Modified Double Base (CMDB) propellants and large size case bonded rocket motors *etc.* He formulated methodology for life extension of propulsion systems of imported missiles, resulting in huge foreign exchange saving. His other noteworthy contributions include development of low vulnerability high energy gun propellants, high performance and thermally stable explosives and explosive reactive armor (ERA).

Dr Singh has been the recipient of a number of awards. He received *Great Indian Achiever Award* (2004) for setting high order of excellence in the field of HEM; *AGNI Award* for Excellence in Self Reliance (1998) and the *Astronautical Society of India (ASI) Award* (1994) in the field of rocket and allied technologies. In recognition of his outstanding Scientific contributions, he was conferred the *Scientist of the Year Award* twice (1993 and 1983). In addition, he is the recipient of the *DRDO Cash Award* and *Best Paper Award* of the Defense Science Journal (1984). Dr Singh has been awarded *Rising Personalities of India Award* for his outstanding services, achievements and contributions. He has been conferred 'Samaj Shri' by Indian Council of Management. He has received the prestigious *Technology Leadership Award* recently for his outstanding contributions in the field of solid rocket propellants, commercial explosives, high-energy materials and armaments.

Dr Singh was the Director of High Energy Materials Research Laboratory for 14 years (1990–2004). He was elevated as outstanding scientist with effect from July 2001. He worked as 'Emeritus Scientist' (DRDO) at ARDE, Pune from 2004 to 2006. Dr Singh worked as a 'Visiting Professor' at Israel Institute of Technology (IIT) Haifa during 2004–2005 and 2009. He was the Chairman of "HEM" research panel and Armament Research Board (ARMREB). Now, he is working as senior technical advisor to private companies engaged in the area of civil explosives, applied chemistry *etc.*

Dr Himanshu Shekhar
Scientist
High Energy Materials Research Lab,
Sutarwadi, Pune – 411021

Born in 1972, Shri Shekhar was adjudged first in his batch during graduation in Mechanical Engineering. He scored 99.57 in GATE 91 and completed his M.Tech in Mechanical Engineering from IIT, Kanpur in 1993 with cent percent CPI *i.e.* 10.00 out of 10.00. He joined DRDO as Sc'B' and is now working as scientist 'F' (Joint Director) at its premiere research lab "High Energy Materials Research Lab" popularly known as HEMRL, Pune.

He is a team leader of a group entrusted with the task of augmenting rocket propellant processing facilities for development of advanced solid propellants. This task involved design, development, fabrication, erection, commissioning and testing of sophisticated equipment like Abrasive blasting and Abrading machine, Degreasing machine, Liner coating machine, Double cone blender, End Trimming machine, Remote controlled fire extinguishing system *etc.* He has been instrumental in preparation of specifications, realization and installation of numerous other facilities like horizontal and vertical curing ovens, mixer bowls, star shaped cores, vacuum casting chamber *etc.* He has designed an import substitute liner slurry-coating nozzle for the plant. As in-charge of facilities, he has processed more than 30 tons of propellants and ensured micro-event processing quick, smooth and safe. He has contributed for structural integrity analysis of propellants, performance prediction of solid rocket propellants, process modeling for solid rocket propellants.

His major scientific contributions include modeling of liner slurry spray coating process, propellant casting under vacuum and transient heat transfer during curing. Shekhar has more than 118 technical publications in international and national journals and has authored nine books. He has won first prizes for *Science Day Competitions* twice and *Technology Day Awards* three times. In addition, his work has been recognized by conferring *Mr Engineer Award* (2003) by the institution of engineers (India), *National Science Day Oration Awards* 2003 (DRDO), *AGNI Awards for Excellence in Self-Reliance* 2001. He was awarded *Young Scientist Award* 2004 for his meritorious contributions

in the field of modeling the process parameters and performance prediction schemes for propellants and high explosives and indigenous
development of automated critical machineries and components for
processing case bonded solid rocket propellants. He is a life member
of High Energy Materials Society of India (HEMSI) and Aeronautical
Society of India (AeSI).

Shri Shekhar has a flare for writing poetry in Hindi; a collection of
his poems has been published in the book entitled "Tathya-Tarang",
which has been conferred with Shri Ram Naresh Tripathi Shikhar
Sahitya Samman award. He is a badminton player and has represented
HEMRL. His other book in Hindi "Raket Rahasya" received the DRDO
Rajbhasa Pustak Award in 2010.

Contents

Solid Rocket Propellants: Science and Technology Challenges
By Haridwar Singh and Himanshu Shekhar
© Haridwar Singh and Himanshu Shekhar 2017
Published by the Royal Society of Chemistry, www.rsc.org

1

The History of Rocketry and The Systems Involved

1.1 History of Rocketry

Fire is the origin of weapon development in a true sense. The throwing of fire pots, containing flammable materials like naphtha, is reported as far back as 1000 B.C. Although not rockets in a true sense, Archytas, a Greek philosopher, demonstrated the reaction principle in 360 B.C. He filled water in a hollow clay pigeon and set it over fire. The pigeon moved under its own power due to the escape of steam through strategically placed holes. In the first century AD, Hero from Alexandria demonstrated the reaction principle using an aelopile, in which a globe mounted on two central trunions rotates due to passage through tangentially placed exit points. The book by Sir Issac Newton *Philosophiae Naturalis Principia Mathematica* (Mathematical Principles of Natural Philosophy) in 1687 resulted in the first scientifically defined reaction principles.

The Chinese were the leaders in the development of firearms. They contributed immensely to both the theoretical and practical development of rocketry. By 200 B.C., the Chinese are believed to have discovered black-powder, while separating gold from silver during a low temperature reaction. They added KNO_3 and sulfur to gold ore but forgot to add charcoal. They added charcoal as the last step. Unknown then, they had made Black-powder, which resulted in a tremendous

Solid Rocket Propellants: Science and Technology Challenges
By Haridwar Singh and Himanshu Shekhar
© Haridwar Singh and Himanshu Shekhar 2017
Published by the Royal Society of Chemistry, www.rsc.org

explosion. Black-powder was, however, not introduced until the 13th century. In 994 A.D., the Chinese developed an attack mechanism based on artillery fire made up of catapulted stones and fire arrows launched by bows. In 1045, a compendium by Tseng Kung-Liang named as "Wu-ching Tsung-Yao" (Complete Compendium of Military Classics) was compiled, which illustrates the use of ballistic fire arrows not launched by bows but by charges of gunpowder. These fire arrows were propelled by ignited gunpowder housed in tube tied to the arrow. These fire arrows were launched in salvos from arrays of cylinders or boxes, which could hold as many as 1000 fire arrows each. In 1500, the Chinese even attempted to propel man with the help of a similar rocket-propelled vehicle, but failed.

The word "Rochetta", which means "rocket" in English, was used first by an Italian named Muratori in 1379 by the 13th century, the armies of Japan, Korea and India are believed to have acquired a sufficient knowledge of gunpowder-propelled fire arrows. In 1285, the Arabs began using gunpowder-propelled fire arrows in combat. By 1410, briefs on the design of tube-launched military rockets were also published. During the 15th century, French troops used war rockets extensively in their attacks. In 1627, gunpowder was used as a blasting agent for recovering ore in Hungary. During 1670, the British used black powder for copper mining. Indian troops were not far behind in using rockets in battlefields. In 1788, Hyder Ali formed a rocket contingent made up of 1200 men. His son Tipu Sultan used it effectively against the British army in the Battle of Srirangapattam in 1792. The rockets disoriented the British soldiers by sheer numbers, sound and dazzling blue light, even during the night.

In 1804, William Congreve developed a variety of superior rockets with incendiary effects, conical metallic warheads, parachute mounted flares and battlefield messages distribution services. He developed a variety of rockets of different calibers, types and for various purposes. Congreve rockets were successfully used in battles for capturing Callao (1809), Cadiz (1810), Leipzig (1813), Fort McHenry in Baltimore (1814) and Waterloo (1815). In the middle of the 19th century, William Hale developed spin-stabilized rockets for better accuracy. The first use of these rockets took place in Mexican war of 1846–48. Russia test fired rockets in 1817 and their first rocket manufacturing plant was established at St. Petersburg in 1826. In the latter half of the 19th century, significant advances in conventional artillery resulted in the reduced use of rockets in wars.

In 1855, the first two-stage rocket was developed for the transport of heavier cord and in rescue line applications. In 1881, Russian Nikolai

Kibalchich is believed to have designed the first rocket-propelled aircraft and the first gimbaled engine. Although unconfirmed, but Peruvian chemical engineer Pedro A. Paulet proposed the first liquid-fueled rocket; he used nitrogen peroxide and gasoline for propulsion. However, by the middle of the 19th century, the limitation of black-powder as a blasting explosive became apparent. In 1846, the Italian professor Sobrero discovered liquid nitroglycerin (NG). A few years later the Swedish inventor, Nobel developed a process for manufacturing NG. Nobel began to license the construction of NG plants, which were built near the site of intended use, as transportation of NG tended to generate a loss of life and property. The Nobel family suffered many setbacks in marketing NG. One of the accidental explosions destroyed the Nobel factory in 1864 and killed Alfred's brother Emil. After another explosion in 1866, which demolished another NG factory, Alfred turned his attention to safety issues for transporting NG. Alfred mixed NG with "Kieselguhr". This mixture was known as guhr dynamite and was patented in 1867.

Along with NG, the nitration of cellulose to produce nitrocellulose (NC) was being studied by different workers. With the announcement of Schonbem in 1846 (and by Bottger) that NC had been prepared, its utilization began. Many accidents took place during the preparation of NC and many plants were destroyed in France, England and Austria. Abel (1865) showed that through the process of pulping, boiling and washing, the stability of NC could be greatly improved. In 1875, Alfred Nobel discovered that on mixing NC with NG, a gel was formed. This gel was used to produce blasting gelatine. Later in 1888, ballistite, the first smokeless powder consisting of NC, NG, Benzene and camphor was discovered. The British called it "Cordite". In various forms cordite remained the main propellant of British force until 1930. The British established a cordite factory in India, close to Otty (Arvankadu) to manufacture various types of cordites.

Before WWII, propellants were established mainly for small arms in cannons and for sporting ammunition for civilian uses. The advent of rockets in WWII and the use of extruded double base propellants in rockets as early as the 13th century by the name of "Fire Arrow" propelled by rockets increased their range. In the latter part of the 19th century, the development of artillery with a high accuracy and long range was due to improvements in propellant characteristics, which continue till now. The two main classes of propellants, solid and liquid have some characteristics in common but there are many more that are quite different. Today, hybrid rockets using a liquid oxidizer and a solid fuel, are gaining importance with respect to their high performance and high safety level.

Although a number of improvements in gunpowder were made, it still had many undesirable properties like bright muzzle flash, a large quantity of smoke and hygroscopicity. The solid residue formed was also very corrosive and had to be removed after each firing. The introduction of NC smokeless powder by Vielle in 1886 marked a significant advancement in propellant history. After a few years Nobel introduced NC–NG based double base propellant. As a result of increased research and development activities during WWII, a group of solid propellants called composite propellants emerged. The first composite rocket appeared somewhere during 1945. Since then composite propellants have assumed a major role in the propellant field. Earlier propellants were used mostly for military applications, but the advent of sputnik and explorer satellites opened the way for greater usage of propellants in space. There has been considerable propellant applications for industrial use but when compared to military use, these applications are limited. Oil-well perforating guns, industrial cannons for quarries are a few examples of industrial use. Jet-assisted take-off (JATO) rockets have been introduced for aircrafts.

In the early 20th century, the preparedness for world wars resulted in the introduction of several new technologies in rocketry. The first guided missiles were introduced in the form of the British A.T. (Aerial Target) and the US Kettering Bug. The feasibility of radio guidance was established during this project. The Kettering Bug, a bi-plane bomber, was successfully demonstrated in 1918. However, these could not be placed into production due to the end of the hostilities of WWI. In between the two world wars, the development and evaluation of several rockets, based on solid and liquid fuels continued in different countries, *e.g.* Larynx, a radio guided mono plane, Queen Bee and Queen Wasp, both radio guided bi-planes (British), GIRD-X and Aviavnito (Russia), Mirak-I (Minimum Rocket–I), Huckel-Winkler-I (powered by liquid oxygen and liquid methane), Repulsor of Germany *etc.* In fact, major development in this vital field took place in Germany after the establishment of a rocket production facility at Peenemunde and production of the 'V' series of rockets. JATO, based on solid and liquid fuels, was developed and introduced by the US During WWII, the US introduced the Bazooka (a rocket-powdered grenade), the Barrage rocket (an air-to-surface missile with many variants like: M-8, super 4.5-incher, spinner, HVSR or high velocity spin-stabilized rocket, Tiny Tim, Bat, T-22, Little Joe, Lark *etc.*). Parallel to this, the British rocket development program included a finned version of barrage rockets, Snare and winged missile "Stooge". The Russians deployed their barrage rocket named "Katyusha".

The Japanese developed the surface-to-air missiles Funryu-2 and 4 and the solid propellant-propelled suicide plane "Ohka" during this period. After WWII, the allied forces captured the Peenemunde plant of Germany and further development of rockets were mainly offshoots of the technical expertise and knowledge gained due to the concentrated efforts of Von Braun of Germany.

Conventional propulsion systems based on gasoline or jet fuel need atmospheric air for their operation. However, rocket or gun propellants do not need air because the required oxygen is contained within the propellants. The propellant system is a balanced source of potential energy containing oxidizer and fuel for combustion or conversion to kinetic energy. When the propellant has an oxidizer and a fuel in one molecule, like NC or nitromethane, it is called a "mono-propellant". However, if the fuel and the oxidizer remain separate and are then mixed in a combustion chamber, they are called "Bi-propellant" systems. Composite propellants have their fuel and oxidizer in separate solid phases. The solid propellants can be broadly classified into three separate groups: (1) homogeneous propellants, containing NC, NG, stabilizer *etc.* also known as double base propellant (DBP); (2) composite propellants (CP) consist of an oxidizer (AP), a binder (polymeric material) and a metallic fuel (Al); and (3) composite-modified double base (CMDB) propellants take advantages of both the DBP and CP systems.

Solid propellants undergo decomposition by a deflagration process. Sufficient heat is generated above the propellant surface, which is transferred back to the surface by conduction and thereby causes further decomposition of the newly exposed surface. This reaction is self-propagating. Propellants perform their work by the slow liberation of energy characterized by high temperature gases pushing against the surrounding air. High explosives, on the contrary, perform their work by sudden shattering, as in the case of rock breaking. Solid propellants were used in the early rockets and have always been used in guns.

1.2 Spacecrafts and Rockets

The development of propulsion units for spacecrafts has been inseparable from the development of rockets. Although not starting until the mid-20th century, space vehicle development has in fact surpassed the wartime uses of rockets. The most popular among US space vehicles is the space shuttle approved officially in 1971 as a

"Space Transportation System" (STS). The argument put forth during the sanction about its utility was to ferry people and supplies; to act as orbiting scientific laboratory; and to place, repair and recover satellites in orbit. Its first flight took place in 1981. For the upper stage launch system, Centaur and Agena have been developed. Vanguard, started in 1955, placed the first satellite in orbit in 1958 only after several unsuccessful attempts. The Titan rocket was developed by the US as a powerhouse for an ICBM (Intercontinental Ballistic Missile). Titan-II, developed in 1962 as a 2-stage vehicle, was successfully produced. However, obsolescence resulted in diversion of these vehicles for space applications as launch vehicles. Later the Titan-III series and Titan-IV were also developed for specific applications. A similar history surrounds the development of "Thor", an intermediate range ballistic missile. The US developed it for deployment in England, in 3-stage and 4-stage configurations. Deployment started in 1958 but later on it was diverted to space application and during 1962, it was used for high-altitude tests. Several versions of hybrid launch vehicles by combining different stages of Thor with stages of Vanguard, Delta, Agena were made and used for space applications. Solid fuel rockets "Scout", instrumental in successful launch of Explorer-9 in February 1961 was grounded by NASA (National Aeronautics and Space Administration) in 1994 after around 118 flights. The development of the Saturn V class of launch vehicles resulted in the successful lunar mission. The 6th Saturn-V propelled Apollo-11 for the first landing of humans on the Moon surface on 20 July 1969. Furthermore, the Saturn class of vehicles was used for the launch of the manned Skylab during 1972–73. The first American satellite, Explorer-I used a Jupiter-C rocket, which was conceived by adding a fourth stage to the Redstone rocket, first launched in 1961. A highly successful but lesser known rocket developed by the US was Delta, which had several feathers to its caps like the launch of the first communications satellite to Earth orbit, the launch of the first geostationary satellite, the launch of the first Intelsat (International Telecommunications Satellite Organization) besides the launch of the Explorer satellites, pioneer interplanetary probes and most of the satellites in the TIROS and Landset series.

A separate chronological look at the launch of satellites is a true depiction of the development and progress of launch vehicles. The Russians launched the first Earth orbiter (Sputnik-I) in 1957 and the first animal in space went in Sputnik-II in the same year. Luna-I, II and III were launched in 1959 to land and take photographs on the moon surface. In 1961, the first manned spaceflight took place in Vostok-I with Yuri Gagarin. The first British-built satellite (Ariel) was launched by the US in 1962. Humans landed on moon in 1969, collecting lunar samples. The first Japanese Earth orbiter (Lambda) was launched in 1970 and the same year the Chinese and French also attained success in sending their orbiters. In 1975, the Venera-9 lander from the Soviet Union reached the surface of Venus. Voyager-I of the US passed by Saturn in 1980 and sent spectacular pictures. In 1983, Pioneer-10 became the first probe to venture into interstellar space, when it crossed Neptune on 13th June. However, it slowed down later and Voyager-I passed Pioneer-10 on 17th February, 1998 to become the most distant human-made object in space. Voyager-I and Voyager-II crossed terminal shock on 15th December 2004 and 5th September 2007, respectively. On 25th August 2012, Voyager-I entered interstellar space, a remarkable achievement for humankind.

With the launch of Sputnik and Explorer, we are now in a space age. What was pure fantasy is now getting serious attention by scientists and technologists. Each day there are new speculations about spacecrafts, particularly after the successful landing on the moon. Oberth published the theoretical basis for space flight in 1923. In 1944, high altitude research studies were launched in the US. An altitude of 65 km was reached in 1945 with a rocket. In 1946, the German V2 rocket reached an altitude of 170 km. In 1952, the school of aviation medicine sent rats and monkeys to an altitude of 60 km. The space programs got a big boost in view of our curiosity about unexplored space and a desire to go where no one has gone, so far. If space is to be used for military purposes, countries of the world should be prepared to protect themselves. A strong and advanced space technology shall place the concerned nation as world leader. Moreover, space missions will help in advancing our knowledge about our solar system, universe and also about our own planet.

Spacecrafts (and rockets) use solid, liquid and hybrid propellants (liquid oxidizer, solid fuel) for propulsion. DBP use NC/NG as the main ingredients. They are also referred to as "Colloidal Propellants". A typical single base (NC-based) propellant used in gun powder consists of NC – 60–99%, Diphenylamine (DPA) – 0.2%, K_2SO_4 – 1–3% and candelila wax – 1–2%. They are thermoplastic in nature, which mean they soften at high temperature and have a waxy appearance. Compositions made by a solvent-less processes are hard and horn-shaped. Numerous shapes like cruciform, tube, rod and tube, internal star, multi-perforated *etc.* have been used for rocket application. Single base, double base and triple base (NC, NG, nitroglycerin or picrite) propellants have been used for gun propulsion. Double base propellants are processed into desired shape by either extrusion technique or by casting process. The casting process uses small granular (1 mm × 1 mm) casting powder. These granules are poured into a mould (Al, steel) of proper size with mandrel (core) to obtain desired shape and then desensitized NG is added from the bottom under vacuum. Then cast propellants are cured and mandrel is extracted.

Composite propellants (CP) are a mixture of finely granular oxidizer, mainly ammonium perchlorate (AP), and a binder (polyurethane, HTPB, CTPB). Since the oxidizer is the major constituent of a composite propellant, it contributes most to the burning process. The various oxidizers used include ammonium nitrate (AN), potassium perchlorate (KP), and ammonium perchlorate (AP) *etc.* For binders starting from asphalt, phenolic resin, synthetic rubbers *etc.*, the modern hydroxyl terminated polybutadiene (HTPB) have been used. When elastomeric binders are used, flexible grain results, which can be directly cast into motor casing. Such propellants are called "case bonded rocket propellants". A typical composite propellant contains 60–75% oxidizer (AP), 15–20% binder (HTPB), 15–18% metallic fuel (Al) along with additives (5%) to impart desired properties to the propellant.

Liquid propellant based rockets differ from solid rockets in that they need a combustion chamber and a feeding system for propellants from storage tanks. Liquid propellants have been extensively used for missile and spacecraft propulsion. Figure 1.1 describes the basic features of solid and liquid propelled rockets.

Both mono and bi propellants can be used. Typical examples of monopropellants include nitromethane, ethylene oxide, ethyl nitrate *etc.*

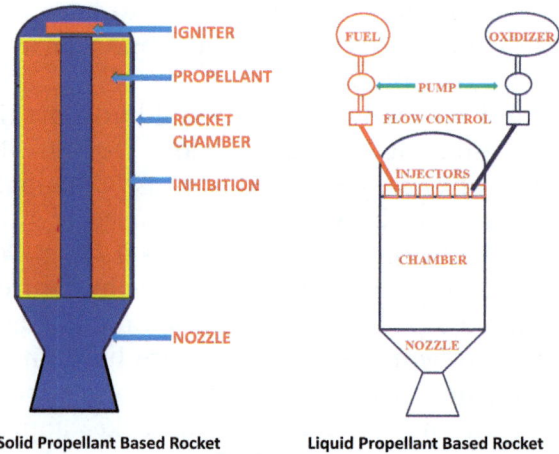

Solid Propellant Based Rocket   **Liquid Propellant Based Rocket**

Figure 1.1   Solid and liquid propelled rockets.

In bipropellant systems, the fuel and oxidizer are stored separately and mixing takes place only in the combustion chamber. Self-igniting or spontaneously igniting bipropellants are called "hypergolic". Those that need an external source of ignition, are called non-hypergolic propellants. The various oxidizers used include HNO_3, liquid oxygen, N_2O_4, H_2O_2 etc. The various fuels used are alcohols, gasoline, hydrazine, *etc.*

1.3 Systems Involved in Rockets and Missiles

Any rocket motor has five major components namely: (i) rocket motor casing, (ii) propellant, (iii) ignition system, (iv) inhibitor, insulator and liner, and (v) nozzle.

1.3.1 Rocket Motor Casing

The casing material of rockets has to withstand both high pressures and the hot combustion gases generated by the burning of the propellants. The strength of the casing material at high temperatures is an important criteria for material selection. In the cartridge loaded mode type system, where propellants are separate from the casing and are loaded in the casing like bullets of a gun projectile, the rocket motor casing is painted with a heat resistant coating to restrict the rise in temperature of the casing. In case-bonded mode type systems, a thermal insulation layer is pasted at the inner surface of the casing to restrict the temperature of the casing to 100–150 °C. AISI4300, Ladish D6 low alloy steel, Maraging Steel, 15CDV6 are the mostly used casing materials for rockets and missiles. Another criterion is the search for a low-density material, so that the mass fraction of the propellant can be increased. This includes composite materials like Kevlar, resin impregnated glass fiber *etc.* With development of better resins and fiber strands, the future of casing materials can be made stronger and stiffer to produce a more efficient pressure vessel (defined by pressure–volume product per unit weight of case). For high energy, highly loaded fully stress-relieved propellants, development work has been initiated to develop the technology for wrapping the motor case over propellant grain, nozzle and ignition system. For repeated use, especially in the case of launch vehicles, material selection criteria includes water impact load bearing capacity and corrosion resistance.

1.3.2 Propellants

Propellants are the power behind rockets, missiles and launch vehicles. They are energetic materials that are ejected as the hot gaseous products of combustion from the nozzle to produce forward thrust. They can be liquid, solid or gaseous. Fluids as propellants have complications during storage, actuation (feeding, piping, valves) and environmental exposures. However, liquids produce specific impulse of the order of 400 s, which is higher than solid propellants. The perpetual search for ingredients of solid propellants has resulted in improved specific impulse, density, burn rate and mechanical properties. But the upper limit of delivered specific impulse for solids has to cross the 300 s barrier to match the performance of liquid and hybrid propulsion systems. Efforts are also being made to improve the mechanical properties in addition to achieving higher performance. This needs higher solid loading, resulting in susceptibility to structural failure, granulation and deflagration to detonation transition (DDT). A very low binder content leads to extensive fracturing of propellants and burning front proceeds towards detonation waves. Hydroxy terminated polybutadiene (HTPB) binder based propellants with AP oxidizer and high solid loading (86–89%) have higher performance, mechanical properties and lower thermal coefficient of expansion. Another requirement put forth for propellants used in tactical missiles is to produce minimum exhaust smoke and low IR signature of the exhaust product of propellant combustion. This ultimately needs reduced aluminium content or non-aluminized propellants, good physical properties over a wide temperature band and high safety. These issues are being debated for achieving better high energy non-smoky propellants. Ageing characteristics and reproducibility and reduced hazard and sensitivity parameters are assuming high importance.

1.3.3 Ignition System

The initiation of propellants needs external heat flux, which is given by pyrotechnic or pyrogen igniters. In pyrotechnic igniters, currently boron-potassium nitrate based pellets and powders are filled in a container. This is initiated by a squib using electric discharge, which produces hot combustion products for initial pressurization of the propellant port and bringing the surface temperature of the propellant to auto-ignition temperature to establish self-sustained combustion. The quantities of igniter composition and container design are important factors for an efficient pyrotechnic igniter. The ignition delay, rate of rise of pressure and initial pressure peaks are the signals, indicating adequacy of ignition system. Pyrogen igniters are used in large size motors and are itself a small motor initiated by pyrotechnic compositions. Safety and arming devices for ignition systems becomes a prime requirement for reliability and survivability. To reduce EMI (electromagnetic interference) and stray voltage from ground loops, inductive coupling and short circuit, air-borne systems need solid-state switching devices with multiple electric interlocks. The firing circuit for igniters may shift to remote, low voltage (28 V dc), capacitor-discharge hot bridge wire devices. Capacitor discharge needs smaller batteries, no high-voltage components, minimum packaging space and no heat sink requirements. Laser beam based initiation systems are also being studied.

1.3.4 Inhibition, Insulation and Liner

There are certain non-energetic, inert (but essential) components in rocket motors. For the cartridge loaded class of motors, propellant grains are inhibited to selectively restrict the burning surface and thereby getting desired thrust time profiles for a given configuration. There are many methods of inhibiting solid propellants like casting, tape winding, cloth winding, brush coating *etc.* The inner surface of cartridge loaded rocket motor's combustion chamber is also coated with heat resistant paint. For case-bonded solid propellant motors, a thermal insulation layer lies between the propellant and motor casing. This prevents the rise in skin temperature of the casing, acts as a cushion to allow the propellant to withstand various handling and flight loads and also absorbs the heat of combustion gases by endothermic pyrolysis. A very good interface strength is always desired for case-bonded motors in case of casing–insulation and insulation–propellant contact zones. For better adherence, a thin layer of liner exists at the propellant–insulation interface. The liner material prevents plasticizer migration also. In the case of case-bonded motors, inhibition lies at the ends of the propellant grains. These inert materials are a must but are also undesirable for the propulsion system's performance. They add extra weight and are non-energetic in nature. Therefore, improvements in ablation rate and a reduction in density are two major research areas in this field.

1.3.5 Nozzle

The nozzle gives way to the high pressure propellant combustion gases, which are discharged at a high velocity in a rearward direction to produce forward thrust to the rockets and missiles by reaction forces. An optimum expansion ratio of combustion gases to the outside pressure becomes a prime requirement and effective exhaust velocity is considered a true measure of the propulsion system performance. Two extreme ends in the design of nozzles are being worked out. A simple nozzle has a convergent and divergent portion in which exhaust gases expand. However, extendable exit cones for strategic motor nozzles are complex in design and operation. The selection of nozzle material, especially throat inserts, becomes significant due to the fact that very high temperatures, coupled with very high heat transfer, exists. Carbon–carbon and graphite–graphite families of composite materials have been an ideal choice for nozzle material. High modulus, increased density (2 g cm^{-3}) and pyrolytic graphite with new orientation are technology improvements in nozzle materials. Nozzle performance is also improved by operating the motor at higher chamber pressure, high expansion ratios and reduced erosion. Two-phase flow losses can be further reduced by short-sharp throat and reduced exit-cone particle impingement. Instead of having conical nozzles, bell shaped nozzles are reported to be efficient, although are difficult to manufacture.

The components are shown in Figure 1.2. The volume fraction occupied by various components in a typical rocket is also given in Table 1.1.

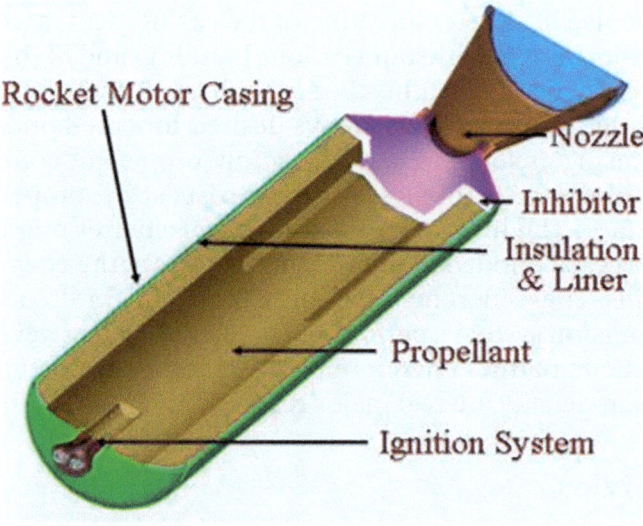

Figure 1.2   Components of rockets/missiles.

Table 1.1   Volume fraction occupied by various components in a typical rocket.

Motor component	Fraction
Propellant grain	75.5
Insulation	1.5
Grain supports	0.6
Igniter	0.4
Free space	21.8

1.4   Other Propulsion Systems

Although chemical propulsion has been the power behind almost all rockets, missiles and launch vehicles, several other concepts are also being investigated for their potential use in propulsion. There are several ways in which higher performance can be obtained but at the same time getting high thrust has been a major challenge. Propulsion system performance has been conventionally expressed in terms of thrust delivered by unit mass flow rate, popularly known as specific impulse and is expressed in seconds (s). Solid propellants are capable of developing specific impulse (Isp) of 250 s, while liquids can deliver up to 400 s (Cryo). Fuel-rich propellants can deliver Isp of more than 1200 s in a secondary chamber. Various options for propulsion are discussed in this section.

1.4.1   Nuclear Propulsion

Energy of the order of 10^6 to 10^8 times than that created by chemical reaction, is released by splitting (or fission) of a nucleus. It is feasible to utilize this controlled energy by means of a neutron-induced fission chain reaction in U^{235}. The second type of nuclear reaction, thermo-nuclear or fusion involves the release of energy from joining (or fusing) the charged nuclei of light elements. The reaction results in the formation of new elements with mass equal to the sum of masses of the reacting elements. The problem of controlling and containing such high temperatures is a challenge before propulsion engineers use nuclear energy.

The most likely method of using fusion for propulsion are the production of thrust by thermodynamic expansion of hot reaction products and the use of a heat transfer system for conventional propulsion, such as heated or ionic rocket propulsion. By adopting the principle of cathode ray tubes, an ion rocket could be created.

The thrust would be generated from the ions and electrons, which are accelerated to velocities, approaching the speed of light. Although specific impulses are extremely high (millions of seconds), the total thrust is small and would only be suitable for outer space, away from gravitational forces.

Photon propulsion is the most promising from a performance point-of-view. However, it is technically very difficult. Particles of finite mass can not achieve the velocity according to the theory of relativity. Both fusion and fission can provide an excellent source of high-energy photons. However, the problem of lining up these high-energy photons is an insurmountable task, currently.

When free radicals recombine, they release enormous amounts of energy. The basic problem associated with free radical propulsion is to formulate, stabilize and evaluate free radicals. If a mixture of 75% hydrogen free radicals is used as a working fluid of hydrogen, a Isp of more than 1000 s would be observed. Furthermore, an extreme temperature will be created from the fuel consisting of atomic hydrogen combustion and the chamber may evaporate at a very high temperature of 4000 °C; considerable re-dissociation may also occur.

In nuclear propulsion, criticality lies with thermal resistance of reactor material and methods adopted for conservation of neutrons for self-propagating fission. In initial stages, graphite has been used as the moderator, but this resulted in the formation of acetylene and other gaseous compounds. So proper protective coating over graphite becomes an essential requirement. In chemical propulsion, the highest achievable exhaust velocity is 4000 m s^{-1}, whereas in nuclear propulsion exhaust velocity as high as 7500 m s^{-1} is possible. Since nuclear rockets needs shielding, they are very heavy and are never preferable for smaller missions.

There are several variants for nuclear propulsion. Fluidized suspended particle bed type reactors have solid fuel particles moving in the stream of high velocity propellant gas. This results in very good heat transfer, small time in solid gas boundary layer and potentially high power density. Fuels used in this case include uranium carbide/niobium carbide, uranium carbide/zirconium carbide or uranium carbide/hafnium carbide and Isp obtained is around 1100 s. In liquid core reactors, gas bubbles through the central core of molten fuel are used for efficient heat transfer but loss of fuel and limitations of gas flow rate restricts the thrust level possible by these reactors. This system can deliver specific impulse of the order of 1200–1400 s. Similarly, gas core reactors can deliver around 2500 s of Isp. With the introduction of pulsed-rocket firing, the pulsed-nuclear propulsion concept is also presented for researchers, where heated propellant charge transfers momentum over a short interaction. The pulse repetition rate may vary from 1 s^{-1} to 1000 s^{-1} and a Isp of the order of 1800 to 2500 s is possible by this method.

1.4.2 Electric Propulsion

Electric propulsion has been attempted for accelerating exhaust jets. For comparison, acceleration of a single charged positive ion through a potential drop of only one volt would correspond to a chemical rocket combustion temperature of 11 600 K (3 times higher). However, power plant weight offsets gain in the specific impulse and heaviness reduces the spacecraft acceleration to negligible (10^{-4} g). Several variants are in active consideration. In electrothermal thrusters, propellant gases are passed through electrothermal arc jet, struck between two electrodes. Energized gases expand through aerodynamic nozzles and generate thrust. Lighter gases are preferred in this case and ionization of gases has to be avoided for preventing efficiency loss. Research indicates that a propellant with high density and low vapor pressure are superior in performance. Hydrazine is found to be a better fuel as compared to hydrogen and ammonia for this application. These systems are capable of generating Isp from 800 to 2000 s. Electromagnetic propulsion utilizes accelerated plasma to generate thrust. The efficiency of conversion of stored energy into the moving plasma can go as high as 30% and very high exhaust velocities can be produced by this mechanism. Electrostatic class of propulsion needs bombardment of ions or surface ionization techniques. In bombardment ion engine, propellant vapors of mercury or cesium are fed to an electron bombardment chamber. Ions are extracted by application of electric charge and further electrons are fed for maintaining neutrality in the chamber. These systems can deliver specific impulse up to 10 000 s. Since ionization efficiency close to 100% can be realized by cesium on tungsten surface, this system also has potential practical application.

1.4.3 ION Propulsion

ION thruster, as ION propulsion systems are commonly known as, is a form of electrical propulsion, where electrical potential difference accelerates ions to generate thrust. Depending on type of electrical energy used for accelerating ions, these thrusters may be classified as either an electrostatic thruster or an electromagnetic thruster. For a typical ion thruster, an input power of 1–7 kW can accelerate ions to exhaust velocity of 20–50 km s^{-1} and generates thrust of 25–250 mN. Although ION propulsion is not useful as main propulsion device, due to the availability of lower thrust and higher weight of the propulsion system, but higher specific impulse makes it a definite choice for auxiliary propulsion systems. This propulsion system may not be very useful for rockets or missiles, employed for defense applications, but spacecraft propulsion have utilized this system of propulsion. ION propulsion have been successfully used in many spacecrafts. The Deep Space 1 spacecraft, powered by an ION thruster, changed velocity by 4300 m s^{-1}, while consuming less than 74 kg of xenon (propellant). The Dawn spacecraft attained velocity rise of the order of 10 000 m s^{-1}. Applications of ion propulsion include control of the orientation and position of orbiting satellites (some satellites have dozens of low-power ion thrusters) and use as a main propulsion engine for low-mass robotic space vehicles (for example Deep Space 1 and Dawn). ION propulsion have high efficiency of 65–80%, but the thrust generated is very low.

A quick comparison of chemical rocket propulsion and ion propulsion can be made. ION propulsion systems have limited thrust density (force per cross-sectional area of the engine). It creates small thrust levels (Deep Space 1's thrust approximately equals the weight of one sheet of paper) compared to conventional chemical rockets, but achieve high specific impulse, or propellant mass efficiency, by accelerating their exhaust to high speed. The power imparted to the exhaust increases with the square of its velocity while thrust increases linearly. Conversely, chemical rockets provide high thrust, but are limited in total impulse by the small amount of energy that can be stored chemically in the propellants. Given the practical weight of suitable power sources, the accelerations given by ion thrusters are frequently less than 1/1000th of standard gravity. However, since they operate as electric (or electrostatic) motors, they convert a greater fraction of input power into kinetic exhaust power. Chemical rockets operate as heat engines; hence Carnot's theorem bounds their possible exhaust velocity. Ion thrust engines are practical only in the vacuum of space and cannot take vehicles through the atmosphere. This is because ion engines do not work in the presence of ions outside the engine. Spacecraft rely on conventional chemical rockets to initially reach orbit, but ION propulsion may be used in many other thruster applications.

1.5 Milestones for the Development of Rockets, Missiles and Space Vehicles in India

Year	Event
1963	US launched first sounding rocket from Thumba Equatorial Rocket Launching Station, followed by 350 US, French, Soviet and British rockets launched between 1963 and 1975
1967	Launch of first Indian-made sounding rocket, Rohini-75
1980	First launch of Satellite Launch Vehicle (SLV-3)
1983	Integrated guided missile development program begins, with participation of more than 60 public and private organizations
1988	First test flight of Prithvi missile
1989	First test flight of Agni missile. Agni is a technology demonstrator. Many advanced technologies proved
1991	First launch of Augmented Satellite Launch Vehicle (ASLV)
1994	• First launch of Polar Satellite Launch Vehicle (PSLV) • Third successful test of the "Agni"
1996	• First test flight of the Prithvi-II surface-to-surface ballistic missile (range 250 km) • Successful launch of the four-stage PSLV and deployment of 1 ton Indian satellite into 500 mile polar orbit • Successful test firing of the Trishul (Trident) surface-to-air missile
1997	Launch of PSLV and deployment of a 1200 kg Indian Remote Sensing Satellite (IRS-1D) into orbit
1999	• Test of the nuclear-capable Agni-II missile • PSLV launched successfully deployed an Indian remote sensing satellite • An unmanned aerial vehicle (UAV) designed to conduct aerial reconnaissance of battlefields tested (Nishant)
2000	India test fired the medium-range Dhanush missile, a naval version of the Prithvi missile
2001	• First test flight of joint Indo-Russian venture supersonic cruise missile "Brahmos" • GSLV (Geo-synchronous Satellite Launch Vehicle) launched
2006	Agni – III, having range of 3000 km test fired successfully on 9th July
2008	Moon Mission (Chandrayan) successfully initiated on 28th October using PSLV-C11
2010	First successful test firing of S200 rocket on 24th January
2012	First test flight of Agni-V by mobile launcher from Chandipur on 19th April
2013	• First flight trial of land-version of Nirbhay Missile on 12th March • Launch of Mars Orbiter Mission (MOM) by ISRO, as first interplanetary mission on 5th November

(continued)

Year	Event
2014	Mars Orbiter Mission reached Mars orbit on 24th September
2015	• First canisterized launch of Agni-V on 31st January
	• First successful flight test of LRSAM or BARAK-8 Missile
2016	ISRO successfully flight tested India's first winged body aerospace vehicle (Reusable Launch Vehicle-Technology Demonstrator) operating in hypersonic flight regime on 23 May
2017	Chandryaan-II using GSLV Mk-III planned to land on Moon
2020	First probe to study sun called Aditya-1 planned

2

Rocket Propellants: Classification and Manufacture

2.1 Introduction

The earlier verdict of great rocket designers, Tsiolkowsky (Russia), Oberth (Germany), Goddard (US) and Esnault Pelterie (France) in favor of liquid propellant rockets have been contradicted by recent developments in the field of solid rocket propellants. Although the solid propellant based project of the US, "Farside" lagged behind the liquid propelled first space booster of the Soviet Union "Sputnik 1" on October 4, 1957 by 18 days, with the passage of time solid propellants have become more prominent for smaller tactical rockets and also for larger sized rocket boosters. The superiority of solid propellants was established by the ambitious missile development program of the US in the 1960s called MINUTEMAN, where solid rocket propellants were proposed to replace the liquid propellant based earlier missiles. Table 2.1 compares the two systems.

The combination of positive aspects of both solid and liquid propellants is incorporated in hybrid propulsion. In general, hybrid systems use a solid fuel and a liquid oxidizer. The performance level (specific impulse – Isp) of hybrid propellants lies between that of solid and liquid propellants.

Solid Rocket Propellants: Science and Technology Challenges
By Haridwar Singh and Himanshu Shekhar
© Haridwar Singh and Himanshu Shekhar 2017
Published by the Royal Society of Chemistry, www.rsc.org

Table 2.1 Comparison of solid and liquid propelled rockets.

	Solid propelled rocket	Liquid propelled rocket
Motor construction and design	Simple construction and design	Complex due to the presence of a fuel tank, pump and valves
Total weight of propulsion system	Less, due to simple design	High, due to peripheral weight
Size of propulsion system	Smaller volume for same energy output	Suitable for larger size rockets
Presence of moving parts	No	Yes (valves)
Deployment time	Ready to use systems	Filling of propellant takes time due to volatile, corrosive and cryogenic nature of fluid
Additional volume for combustion	No. Burning takes place in the combustion chamber	Yes. Combustion and storage chambers are separate
Rocket motor thickness	Thick casing	Thin casing
Fuel preparation	Hazardous due to high sensitivity of ingredients	Easier. Ingredients can be separately stored in tanks and filled in rocket on requirements
Hypergolic ignition	Not applicable	Possible
Start–stop capability	Not available	Simple and easy start–stop capability
Extinction of rocket after ignition	Difficult until consumption of propellants	Possible by stopping fuel supply
Performance level	Less Isp (<300 s)	Higher Isp (>300 s)
Toxicity of exhaust products	High	Low
Fall of un-ignited rocket	Hazardous	Less hazardous
Cost of manufacture	Low	High
Density impulse	High	Low

Broadly, rocket propellants can be classified into four categories:

1. Solid propellants
2. Liquid propellants (Mono & Bi propellant)
3. Hybrid propellants
4. Gelled propellants

Fuel-rich propellants (FRPs) for Rocket Ramjet are of recent origin. As the name indicates they contain a maximum amount of fuel (Metallic or polymer) with the least amount of oxidizer. The oxidizer is added to the propellant only to ensure sustained combustion in the primary chamber. Using atmospheric air, FRPs can produce Isp in the range of 500–1200 s. Boron, aluminium, magnesium, Mg–Al Alloy, titanium, zirconium, and nickel powders appear to be promising fuels. Research work is being concentrated on polymeric FPRs, particularly polycyclopentadiene-based fuels.

2.2 Solid Propellants

Double base propellants conventionally known as homogeneous propellants emerged during the second half of the 19th century and propelled rocket and missiles during the first half of the 20th century. These propellants are based on nitrocellulose (NC) and nitroglycerin (NG) mainly along with some additives like stabilizers, ballistic modifiers, plasticizers, cooling agents, lubricants, opacifiers *etc.* as the propellant is obtained by gelatinization of NC by NG, this class of propellant is also called colloidal propellants. The details about the propellant ingredients are discussed in Chapter 3. The double base propellants are manufactured by extrusion routes and a high rate of production is established at many plants located all over the world. Such extruded double base (EDB) propellants are used in many tactical rockets like 122 mm grade, 57 mm rocket, 68 mm arrow rocket *etc.* However, enhanced hazards restricted the largest extruded size of the propellant and a casting method was developed for double base propellants. Such propellants are called cast double base (CDB) propellants. Size limitations are overcome but production rate slowed due to adaptation of CDB route of propellant processing. Comparison of EDB and CDB propellants is presented in Table 2.2.

The advancement in chemistry gave a boost to the development of composite or heterogeneous propellants during the second half of the 20th century. In this propellant, conventional gelatinization

Table 2.2 Comparison of EDB and CDB propellants.

	EDB propellants	CDB propellants
Size limitation	Yes due to (i) extrusion press cylinder and (ii) press capacity	No as the mould can be of any size
Defect type	Elongated holes	Voids or blow-holes
Machinery requirements	High capacity press	Vacuum chamber
Case-bonding	Not possible	Possible during casting in insulated rocket motor using impregnated cloth
Homogeneity	Better	Relatively poor
Rigidity	Better	Relatively inferior
Density	Higher (1550–1660 kg m^{-3})	Lower (1500–1580 kg m^{-3})
Inert plasticizer content	Low	High
Heat of explosion	700–1100 cal g^{-1}	500–1100 cal g^{-1}
Burning rates	Higher	Relatively lower
Temperature sensitivity coefficient of burning rate	Burning rate not affected by temperature. Near zero value of temperature sensitivity coefficient	Burning rate decreases, if soaking temperature is higher
Width of glass transition zone	1×10^{-4} to 1×10^{5} min	1×10^{-6} to 1 min
Glassy modulus	2000 MPa	1000 MPa
Rubbery modulus	3 MPa	1.5 MPa
Percentage elongation	Same at low and ambient (~4–5%) temperature	High at ambient (~10%) than at cold (~2–3%)
Critical diameter of detonation	2 mm	14 mm

and swelling of macromolecules is completely eliminated. Although double base propellants have a higher strength and they can be easily used for free-standing cartridge loaded rocket configurations, the low density and low energy are placed as major drawbacks for such class of propellants. The name colloidal propellants, which are popular for double base propellants, became no longer valid for the next class of propellants called composite propellants. Most of the new development during this period has been to use composite propellants to replace less energetic double base propellants. In composite

Table 2.3   Comparison of double base and composite propellants.

	Double base propellant (DBP)	Composite propellant (CP)
Type of propellant	Homogenous	Heterogeneous
Discrimination of fuel or oxidizer	NC and NG have both fuel and oxidizer moieties	Fuel: binder and metal; separate oxidizer
Solidification process	Gelling, swelling or physical mixing	Chemical cross-linking
Transparency	Opacifier needed	Opaque propellant
Manufacturing method	Extrusion, casting (mould)	Casting (mould/motor), extrusion
Isp	Low (200–220 s)	High (240–260 s)
Burning rate index	High (~0.5)	Low (~0.3)
Tensile strength	High (140 kg cm^{-2})	Low (15–25 kg cm^{-2})
Hazard classification	1.3	1.3
Health hazards	NG gives headache	TDI is toxic
Density	Low (~1600 kg m^{-3})	High (~1750 kg m^{-3})
Degradation mechanism	NG exudes on ageing	Moisture sensitive
Shelf life	Longer (15–20 years)	Shorter (10–15 years)
Flame temperature	Low (2000–2500 °C)	High (2400–3200 °C)
Exhaust	Non-smoky	Smoky
Solid loading	Moderate	High

propellants, the oxidizer and metallic fuel is loaded in a binder matrix along with various additives and chemically cured using a curing agent to get solid rocket propellants. A comparison of both the propellants is given in Table 2.3.

In addition to conventional double base and composite propellants, composite modified double base (CMDB) propellants have been extensively used for military applications. A double base matrix of NC and NG is loaded with an oxidizer like ammonium perchlorate (AP) or explosive ingredients like nitramines (RDX, HMX) to enhance the performance. Such loaded double base matrix constitutes the CMDB class of propellants, where a gain of around 20–30 s in Isp is possible.

In the case of AP-CMDB propellants, the oxidizer-rich decomposition products of AP shifts the reaction of double base matrix products towards stoichiometry, leading to an increased reaction rate near the surface and high flame temperature. This enhancement in performance is offset after the addition of more than 50% AP to the double base matrix and a maximum Isp of 250 s is obtained. But AP catalyses decomposition of NG and the composition needs the addition of a supplementary stabilizer to neutralize AP driven NG decomposition. The addition of AP makes the composition more sensitive and exhausts polluting. A chlorinated exhaust product produces secondary smoke in a moist atmosphere.

Nitramine based CMDB formulations are reported to be smokeless. They offer higher performance due to higher flame temperature (3280 K) and low mean molecular weight of exhaust products. The enhancement in Isp is attributed to a positive heat of formation of RDX (+14.69 cal mol^{-1}) and HMX (+17.92 cal mol^{-1}). But this class of propellants suffers from two problems; low burn rates and high pressure index values.

The journey of solid propellants development for rocket has not stopped here. Various advanced ingredients are incorporated in the formulations and solid propellants of higher energy are obtained. Using elastomeric binders like viton and Teflon *etc.*, the castable composite propellants can be extruded and thus evolved the extruded composite propellants, which has potential application in power cartridges for aircrafts. This class of propellants have a high density and can be recycled to make them zero wastage propellants. High density impulse is an added advantage of this class of propellants. The incorporation of energetic oxidizers, (like ADN, HNF) energetic binders (like high molecular weight GAP, BAMO, NIMMO), energetic plasticizers (low molecular weight GAP, TMETN, DEGDN) and alternate metallic fuels (like AlH$_3$) have resulted in enhancement in specific impulse or delivered energy from rocket propellants.

In the area of solid rocket propellants, fuel-rich propellants, which are utilized in ramjet applications, must be mentioned. These propellants have very high loading of metallic fuel (~50%), which gives energy (specific impulse) many-folds higher than conventional composite propellants in secondary mode combustion. Zirconium, magnesium, *etc.* are used as major metallic ingredients in such propellants.

2.3 Liquid Propellants

The origin of using liquid propellants can be traced back to 1895, when Pedro E. Paulet from Peru made a rocket chamber of vanadium steel and used nitrogen peroxide and gasoline as the fuel. He ignited it with a spark plug powered by electricity and this device is said to be the predecessor of the modern bipropellant rocket motor. K. E. Tsiolkovsky made another definite attempt in 1903 in St Petersburg, Russia. He deliberated on the use of liquid oxygen and liquid hydrogen based propellants, which is one of the most powerful systems so far. In 1913, Robert Esnault Peterie also advocated the use of liquid propellants for space flight and similar views were expressed by H. Goddard (1919). In fact, Goddard succeeded in making a short flight of his small liquid fuel projectile in 1926. In Germany, Hermann Oberth proposed the use of ethyl alcohol and liquid oxygen as propellants and used the concept of the regeneration principle for the first time. With this humble start, liquid propellant based rockets became very useful for space missions and also for battlefield weapons.

The utilization of liquid propellants has resulted in several critical technology demonstrations. With an increase in size, the upper stages of launch vehicles started using cryogenic liquids along with multiple start capabilities and bipropellant technology with deep throttling and pulse-mode operation. By the 1960s, hydrazine based monopropellants became workhorse liquid fuel for satellite altitude control and station-keeping. A major problem put forth during operation has been propellant storage and feed systems. This resulted in a new concept of insulation for fuel tanks. For controlling the problem of usage in a zero gravity atmosphere, the development of Bladders, screens and other devices to insure positive expulsion of propellants became mandatory.

There are several ways in which liquid propellants can be classified. One method of classification is the boiling point of their fuels or their use. Using this, the two types of liquid fuels are:

(a) Earth storable fuel. The liquid fuels for rockets that can be stored in a standard Earth environment without phase change. They generally produce less Isp as compared to cryogenic fluids.

(b) Cryogenic fuel. The liquid fuels that need very low temperatures for their storage in liquid form. Their boiling points are less than standard atmospheric temperature on Earth. They remain in gaseous form and are stored at high pressure. They are highly corrosive and also highly reactive in nature.

The values of Isp for some liquid fuel combinations are given in the Table 2.4.

Table 2.4 Isp of liquid propellants.

Cryogenic liquid fuel	Isp (s)	Earth storable liquid fuel	Isp (s)
Hydrogen + fluoride	410	Hydrazine + nitrogen tetraoxide	292
Hydrogen + oxygen	391	Methylhydrazine + chlorine trifluoride	284

2.4   Hybrid Propellants

The hybrid propulsion systems are considered the future of rockets and they need both solid fuel and liquid oxidizers. The solid fuel burns differently from combustion in conventional solid propellant rockets. As heat is transmitted to the solid by a suitable ignition system, it vaporizes to generate pressure in the chamber. This is called the primary reaction and the high pressure actuates the low pressure diaphragm seal to pressurize the tank of the liquid oxidizer. The oxidizer flows through an injector and mixes with the fuel rich mixture of the primary reaction. A high performance propulsive force is generated that is proportional to the flow rate of the oxidizer. It delivers a very high density impulse of the order of 700 s. The only difficulty in using a hybrid system is the necessity of pressurizing the oxidizer tank above the pressure of the combustion chamber for positive flow. This needs a cold gas generator propellants or high pressure gas bottles.

The thrust level can be tailored by oxidizer flow control even during flight. This throttling of oxidizer flow enables ignition rise rate control, low thrust at maximum aerodynamic pressure, and minimum thrust imbalance in multiple motor configurations. The combustion process terminates upon cessation of oxidizer flow. The solid grain regresses perpendicular to the flow of gaseous oxygen and not normal to the grain surface as specified for conventional solid propellants. The vaporized fuel rich products mix with pressure fed oxidizer and combustion takes place above the fuel charge rather than on the surface. So voids and cracks in solid propellants, which are the major cause of propellant rejection in solid rocket propellants, have no impact on the performance of the fuel or combustion chamber pressure and will not lead to any catastrophic failure as in the case of solid propellants.

The most important positive aspect in using hybrid propellants is the confinement of the high temperature reaction to a very small area. This area can have an improved insulation and temperature control mechanism and operating the motor at higher temperature definitely leads to higher performance.

Hybrid charges are inert in nature and are safe to handle. Even liquid tank rupture will not lead to an explosion. The liquid oxidizer simply quenches the combustion products and terminates the pressure and thrust. When compared to liquid propulsion units, it needs injection of only one fluid and fluid injection is much simpler and the feeding system is not so complex.

The solid propellant grains of hybrid propulsion have a higher strength and resilience due to a very low oxidizer content. This reduces creep and cracking of propellant grains. The solid propellants can be prepared in the form of thin wafers and can be stored in the combustion chambers in stacks without any bonding between wafer layers. This improves the propellant mass fraction and a higher performance and reduced combustion chamber size is possible. A hybrid system is capable of reduction in propulsion system length by 30–40% thereby increasing payload volume.

The hybrid system has a start–stop capability and firing can be stopped by shutting off the oxidizer tank valve. Of course inertness of fuel charge leads to problems in re-ignition. However, the provision for separate pulses of gaseous ignition system for initiation can rejuvenate the propulsion system, time and again.

The cost of propelling through a hybrid system is relatively less. A substantial saving on cost for storage, ground processing, transportation and launch operations are possible. These motors can be fired, terminated, inspected, evaluated and restarted for additional testing lowering non-recurring development costs.

2.5 Gelled Propellants

Gelled propellants are basically a variant of liquid propellants, where additives make the liquid thixotropic. Gelled propellants remain like thick paint in stationary condition but can flow through valves, pumps and other devices on application of adequate shear stresses. This behavior reduces the chances of spillage during storage and demonstrates very good flow control under shear. In these gels, solid powdered ingredients can be added to enhance the density of the fuel and long term storage without settling or segregation is possible due to the thickening tendency of the liquid on storage. It is safe to handle and the chances of explosion in the fuel tank is greatly reduced.

This system also suffers from certain disadvantages. The addition of gelling agents dilutes the liquid propellant and reduces its energy output. The loading and unloading of propellant is difficult and pipe losses are more because of adhering thick layer at pipe surface. It is highly susceptible to changes in temperature, which leads to changes in viscosity, density and mixture ratio of propellant.

2.6 Manufacture of Propellants

Before getting a brief insight into popular manufacturing techniques for the realization of solid rocket propellants, it is apt to recapitulate the various types of propellants, discussed earlier. This recapitulation helps in understanding and exploring various manufacturing techniques in detail in Chapter 4. The classification of propellant, in summary-form, is given as Figure 2.1.

Liquid and hybrid propellants are individual ingredients and hence are processed individually by conventional well-known methods. However, in the case of solid propellants, the processing technologies generally used are:

- extrusion
- casting (mould and motor casting)
- pressing

Extrusion technology has been extensively used for double base propellants. It is of recent origin for composite propellants. Extrusion needs a high capacity press to extrude propellant charges of different sizes and geometries. Die and pin assembly design also assumes importance here. The various operations involved in the solvent-less extrusion of double base propellants include wet mixing, dewatering, manufacture of paste, kneading, drying, rolling and extrusion. The propellant is kept at ambient temperature for 2–3 weeks for stabilization. The extruded propellants cannot be processed beyond a certain diameter and thus this technique has limitations in terms of the size of propellant. However, the extruded propellants produce higher density and superior mechanical properties. The use of thermoplastic elastomers like viton and Teflon in composite propellants has resulted in the development of extruded composite propellant grains, which results in higher density of propellants. This method is suitable for the bulk production of small diameter propellant charges mainly for seat ejection cartridges.

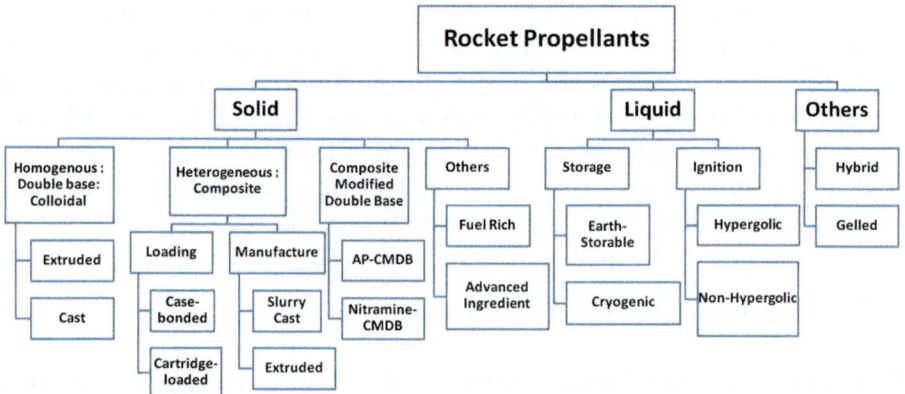

Figure 2.1 Classification of rocket propellants.

For cast double base propellants, the first step is the preparation of the casting powder. The casting powder can be single base, double base or multiple base. The casting liquid consists of a desensitizer, NG and a stabilizer is prepared separately. The casting powder processing is like solvent extrusion of propellant (1 mm × 1 mm). The various operations involved in the processing of CDB propellant include mould filling, evacuation, casting (bottom feeding against gravity), curing, mandrel extraction, machining *etc.* There is no size limitation as such. For cast composite propellants oxidizer and binder preparation, metal powder sieving, drying and incorporation in sigma blade mixer maintained at a certain fixed temperature, casting and curing are main steps. The ingredients are added in installments. After complete mixing is over, curing agents like isocyanates are added to initiate the chemical reaction. The propellant charge is cured at 50–80 °C for 5–30 d, depending on web thickness and type of curing agent.

CMDB propellants and nitramine based propellants can be made by an advanced casting powder (ACP) route or slurry cast technique (SCT), in which spheroidal NC is used. Mixing, casting and curing are like composite propellants. The manufacturing details and flow charts are given in Chapter 4.

Propellant Ingredients and Their Properties

3.1 Solid Propellant Ingredients

The list of ingredients for solid propellants is presented in Table 3.1.

3.1.1 Double Base Propellant (DBP) Ingredients

Homogenous double base propellants (DBP) contain NC and NG as their major ingredients. NC or more accurately cellulose nitrate is prepared from cellulose (waste cotton or cotton linters) using a nitration process. The nitrogen content of NC varies from 12.6–13.5%. Depending on percentage of nitrogen available, NC is classified as type 'A' (12.2% N_2), Pyro (12.6% N_2) and Gun Cotton (GC) with more than 13.1% nitrogen. Stabilisers like 2NDPA or carbamite are added to reduce the degradation of NC. Sometimes two different types of NCs are blended to obtain the desired properties, particularly viscosity. The final product is stored in water-wet conditions. Water is removed at the time of use by treating it with alcohol. In a solvent-less process, the water is not removed until mixing of the other ingredients is complete. A flow diagram for making NC is given in Figure 3.1.

NG is obtained by nitration of glycerol. Sufficient water washes are given to NG to avoid acid contamination, resulting in a gradual degradation of the product.

Solid Rocket Propellants: Science and Technology Challenges
By Haridwar Singh and Himanshu Shekhar
© Haridwar Singh and Himanshu Shekhar 2017
Published by the Royal Society of Chemistry, www.rsc.org

Table 3.1   List of solid propellant ingredients.

Category of ingredients	Name/abbreviation of ingredients
Double base propellants	
Energetic plasticizers	Nitroglycerin (NG), trimethylol ethane trinitrate, (TMETN), triethylene glycol dinitrate (TEGDN), diethylene glycol dinitrate (DEGDN)
Non-energetic plasticizers	Diethyl phthalate (DEP), triacetin (TA), other organic phthalates (DBP, DOP *etc.*)
Binders	Nitrocellulose (NC)
Stabilisers	EC or carbamite (*sym* diethyl diphenyl urea), 2-nitro diphenyl amine (2-NDPA)
Burning rate catalysts	Lead salts (lead stearate, lead salicylate, lead citrate), copper salts (copper stearate and salicylate)
High energy additives	RDX, HMX, NQ (nitroguanidine)
Opacifiers	Carbon black
Flame suppressants	KNO_3, K_2SO_4
Metallic fuels	Al
Combustion instability suppressants	Al, Zr, zirconium carbide (ZrC)
Composite propellants	
Oxidisers	AP, AN, NP, KP, RDX, HMX
Binders	Polyurethane (PU), PBAN, CTPB, HTPB
Plasticisers	DOA, idodecyl pelargonate (IDP), DOP
Metallic fuels	Al, Mg, B, Zr, Ti
Burn rate catalysts	Fe_2O_3, *n*-butyl ferrocene, LiF, oxamide, catocene
Curing agents	Toluene diisocyanate (TDI), isophorone diisocyanate (IPDI), hexamethylene diisocyanate (HMDI)
Bonding agents	Tris-1-(2-methyl) aziridinyl phosphoric oxide (MAPO), triethanol amino (TEA), MT-4 (adduct of 2 moles of MAPO, 0.7 moles adipic acid and 0.3 moles tartaric acid)
Combustion instability suppressants	Al, Zr, ZrC

During storage, NC/NG decompose very slowly (but continuously) generating oxides of nitrogen. Hence, certain chemicals called stabilizers (diphenyl amine, symmetrical diethyl diphenyl urea) are added to slow down the decomposition rate; they cannot stop degradation completely. NO_2 reacts with the stabilizers to produce stable nitrated compounds. 1–2% of stabilizers are generally used.

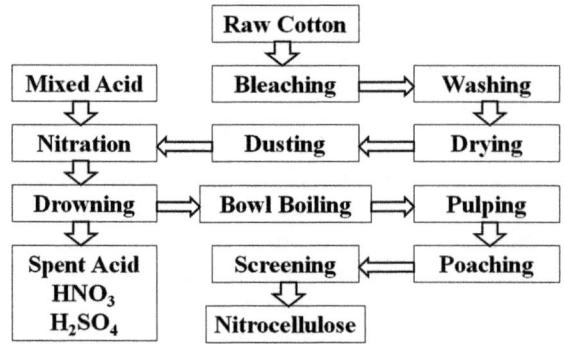

Figure 3.1 The manufacture of NC.

Plasticisers are used to improve the mechanical properties and extrusion characteristics of the propellant. Dimethyl, diethyl and dibutyl phthalates have been used as effective plasticizers (non-explosives). The best explosive plasticizer is NG or diethyl glycol dinitrate (DEGN). DEGN was extensively used during WWII.

To avoid 'worm-holing', the reduction from burning gases, carbon black is used as a darkening agent. It stops radiation transfer and prevents the premature ignition of the propellant by making the propellant opaque; it is therefore also called an opacifier.

Generally, potassium sulfate is added to reduce the flash and smoke of gun propellants. However, in modern DBP, lead and copper salts of organic acids, both aliphatic and aromatic, are used as ballistic modifiers.

Other additives used in DBP include coolants like DNT and TNT. These are used to reduce muzzle flash in gun propellants.

3.1.2 Composite Propellant Ingredients

3.1.2.1 Oxidizer

Oxidizers are a major ingredient of composite propellants (60–80%). An oxidizer is one which possesses a high oxidation potential and a high electronegative atom or group. The periodic classification of the elements can be used to distinguish the oxidizer (high electronegativity) and the fuel (high electropositivity). In other words, the elements on the right side of neutral species like nitrogen in the periodic table are oxidizers whilst those on the left are fuels. Thus, fluorine is the best oxidizer and the F_2/H_2 propellant system gives maximum Isp among chemical propulsion systems. The oxidation potential of the oxidizing group varies in the order $F^- > OF^- > NF_2^- > ClF_4^- > O^- > NO_3^- > ClO_4^- >$

$NO_2^- > ClO_3^-$. It is observed that perchlorates and nitrates are better oxidizers than nitrites and chlorates. The oxidizers used in solid propellants, which are a mixture of an oxidizer and a fuel, must have a high available oxygen content corresponding to a fairly high stoichiometric mixture ratio (W_f/W_o). These oxidizers furnish essentially the oxygen required for the combustion of the fuel binder, releasing the maximum possible energy. The characteristics of oxidizers affect the ballistic and mechanical properties and the processing ability of a composite propellant. Oxidizers are selected to provide the best combination of available oxygen, high density, low heat of formation (low bond energy) and maximum gas volume in reaction with binders. In addition, it should be safe to handle, less hygroscopic, compatible with the other ingredients, easily available, have a good storage life and be low cost. High thermal stability is another desirable characteristic. An increase in the oxidizer content increases the density, the adiabatic flame temperature and the Isp of a propellant to a maximum.

A large number of inorganic compounds have been used as oxidizers. Starting with KNO_3/NH_4NO_3 to $KClO_4$, NH_4ClO_4 and $LiClO_4$. However, phase-stabilized ammonium nitrate (AN) and ammonium perchlorate (AP) are generally used in modern composite propellants. Although the oxidizing potential of AP is less than that of sodium or potassium perchlorates, it produces gaseous products on decomposition and is, therefore, preferred. AP is produced from a sodium perchlorate and ammonia reaction. Alternately, $KClO_3$, used in the match industry, is oxidized to $KClO_4$ and then reacted with ammonia to produce AP.

The workhorse oxidizer used today for composite propellants is AP (NH_4ClO_4). It has a high density and produces the highest practical Isp. It is relatively simple to use, ignites easily, has a reasonable burn rate, does not require a burn rate catalyst and is less hygroscopic. Many AP compositions burn well only under pressure, and the motor must be designed to take this into account. They are best suited for core burning, rather than end burning, motors. AP has an intermediate heat of formation, lower flame temperature (1405 K) and Isp of 157 s. When incorporated in a double base (DB) matrix, it shifts the stoichiometry of combustion products due to a superior oxygen balance (+35%). The oxygen-rich decomposition products of AP interact with the fuel-rich decomposition products of the DB matrix, resulting in an increased reaction rate near the propellant surface leading to higher flame temperature and superior performance. Maximum flame temperature and Isp achieved with 50% AP are 3300 K and 250 s, respectively.

AP has a phase transition at 240 °C (orthorhombic to cubic) and starts decomposing at 439 °C. The decomposition can be catalyzed by metallic salts such as iron oxide and copper chromite at a lower temperature. Very finely divided AP is more sensitive to impact and friction than the coarse material and the presence of hydrocarbons greatly increases the likelihood of a detonable reaction. The burning rate of AP based propellants is also influenced by the particle size of the oxidizer *e.g.* the rate may increase by a factor of six by decreasing the average diameter from 400 μm to 1 μm. Bimodal and even trimodal distributions are used to load the binder with a maximum oxidizer content. The average particle size and particle size distribution of the perchlorate have a negligible effect on the Isp of the propellant. However, composite propellants based on AP produce smoke in cold or humid atmospheres containing large quantities of hydrogen chloride (HCl) as the major exhaust product, which affects the environment. Also military rockets need to have a low signature. This calls for the development of new oxidizers to replace AP.

AN is another oxidizer that is sometimes used in experimental motors. AN based compositions have lower Isp, are less dense and generally require higher solid loading (around 80%) for proper operation. However, an AN propellant still burns more slowly than most AP compositions. Phase-stabilized AN (PSAN), already milled to size is suitable as a propellant.

AN has five crystalline phase transitions (18, 32.2, 84.2 and 125 °C). The phase transformation at 32.2 °C is accompanied by a significant density decrease, causing grain dimension variations. As a result, the propellant grains have a tendency to crack and burn improperly. Efforts being made to shift this phase transition to higher temperatures either by doping or co-crystallization. AN is a high gas producer and is good for gas generator propellants. Furthermore, it does not contain any toxic elements. To boost energies, various oxidizers like AP, HMX *etc.* have been added to the formulation based on AN.

Potassium perchlorate (KP) is used for higher burning rates. It is characterized by high heat output, is a low gas producer and has a specific gravity of 2.5. Nitronium perchlorate (NP), having the dual advantage of positive heat of formation ($+9$ kcal mol^{-1}) and oxygen balance of 66%, is a major contender as an oxidizer. However, it is toxic, decomposes above 80 °C and is very sensitive. Lithium perchlorate ($\rho = 2.42$ g cm^{-3}) is very hygroscopic but can be used for high-temperature applications. However, these compounds could not find practical applications due to their unstable nature and high sensitivity.

New oxidizers that are likely to replace AP in years to come include ammonium dinitramde (ADN), CL-20 (HNIW) and hydrazinium nitroformat (HNF). Their properties are discussed in Chapter 12.

3.1.2.2 Binder

Resinous materials or elastomers that have the capability of binding oxidizer particles are used as fuels. They act as fuel cum binder and their proportion varies between 10 and 25%. Both thermosetting and thermoplastic substances have been used. Binders are typically cross-linked polymers (pre-polymers + crosslinkers) and provide a matrix to bind the solids (oxidizer, fuel, additives) together with a plasticizer to ease the processing of the uncured mix. The major requirements of a binder are as follows:

- It must be a liquid with a workable viscosity (100–10 000 cp) at mixing temperature.
- It should have a molecular weight of 2000–5000, with a minimum density of 0.86 g cm^{-3} for easy processing.
- It must have a reactive functional group (–OH or epoxy) that can be converted to an elastomer by cross linking during curing to obtain good mechanical properties.
- It should have a high heat of formation to be energetic.
- It should have a low glass transition temperature (T_g) to function satisfactorily at extreme temperatures.
- It should have low shrinkage during curing and must be capable of being cured at low temperatures (40–80 °C) with a minimum evolution of heat.

The characteristics of the binder are determined by the following properties of the pre-polymer:

- Average molecular weight—a decrease in molecular weight of the binder reduces viscosity and leads to shrinkage during curing. High molecular weight (chain mobility decreases) renders the uniform mixing of the ingredients difficult and decreases the solid loading capacity. Therefore, the molecular weight of the pre-polymer used as binder applications should be in the range of 2000–5000. The complete curing of the polymer is essential to get the required mechanical properties of the propellant grain.
- Polydispersity—this gives a general idea regarding the distribution of different molecular weight species as well as the lowest and

highest molecular weight species in a polymer system. It is represented as M_w/M_n. Ideally, it should be unity. A narrow molecular weight distribution (polydispersity < 1) also gives lower viscosity for a given average molecular weight. Hence, the pre-polymer should have a polydispersity greater than 1 but less than 2.

- Functionality and crosslinking—the number of reactive functional groups/reactive sites present in a pre-polymer denotes its functionality. The functionality of the pre-polymer should be more than two. Difunctional curatives lead to a linear coupling whereas trifunctional molecules produce three-dimensional crosslinked networks. The ratio of the bifunctional to trifunctional molecules in a binder system must be controlled to provide the required degree of crosslinking and crosslink density. This decides the overall mechanical properties in terms of strength as well as elongation.

Starting with asphalt to poly-butadiene acrylic acid acrylonitrile—terpolymer (PBAN), carboxy terminated poly-butadiene (CTPB) and hydroxy-terminated poly-butadiene (HTPB) have been used as binder cum fuels (Figure 3.2). Polystyrene, polyurethanes, polyvinyl acetate *etc.* were also used as binders for composite propellants.

At present, in most of the operational rockets/missiles and space vehicles, HTPB based propellants are used. There are two popular routes for synthesis of HTPB: free radical polymerization and anionic polymerization. Free radical polymerization is cost-effective and leads to good mechanical properties in the resulting propellant. Anionic polymerization, however, gives a narrow distribution of molecular weight of the pre-polymer. In India, the free radical polymerization method has been perfected using methanol as the solvent, hydrogen peroxide as the initiator and isopropanol as the medium for polymerization. In anionic polymerization, the non-polar solvent toluene gives better results with an organo-lithium initiator. Termination by hydroxyl group is achieved by propylene oxide.

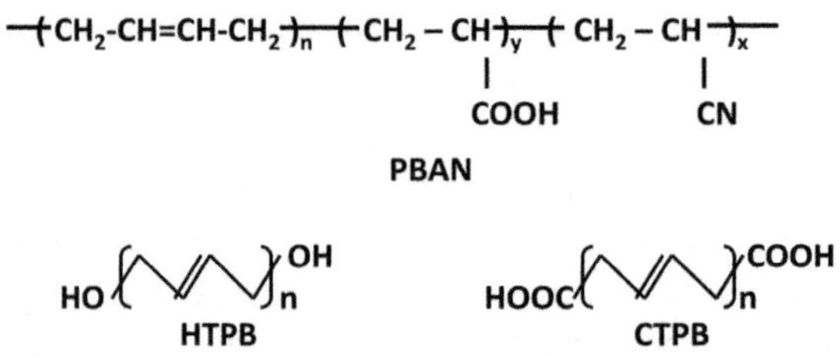

Figure 3.2 Binders for composite propellants.

Incorporation of energetic groups like nitro and azido nitrato in the polymer molecules results in a high density and heat of formation.

3.1.2.3 Plasticizer

Plasticizers are low molecular weight, non-volatile, non-reactive liquid substances. When added to a polymer, they improve flexibility, processing ability and the mechanical properties of the propellant charge. The plasticizers permeate into the polymer chains, thereby reducing the cohesive forces of attraction between the polymer chains and increasing the free volume, which leads to increased chain mobility at a given temperature. This softens the polymer matrix and makes it more flexible.

The first plasticizer used in the propellant industry was nitroglycerin (NG). It is however, highly sensitive to friction and explodes above 200 °C. Other non-energetic plasticizers include dioctyladipate (DOA), dibutyl phthalate (DBP), dioctyl phthalate (DOP) *etc.*

3.1.2.4 Metallic Fuel

Most of the current composite solid propellants contain very fine (10–15 micron) metallic fuels such as aluminium (Al). They increase the chemical energy of the propellant by increasing the combustion temperature (T_f). At low concentrations, Al also helps in promoting stable burning. Boron is another high-energy fuel that is lighter than aluminium but burns with difficulty. Beryllium is highly toxic and therefore cannot be used. Other fuels reported are Ti, Zr and Ni.

3.1.2.5 Other Ingredients

The burn rate requirements are specific to the design of the particular propellant grain configuration to realize the desired thrust–time profile. Common burn rate modifiers used in composite propellants are iron oxide, copper chromite and ferrocene derivatives. These products either accelerate the decomposition of the perchlorate or lower its decomposition temperature. Major ferrocene derivatives used as burn rate accelerators include *n*-butyl ferrocene, di-*n*-butyl ferrocene, catocene, butacene *etc.* Major burn rate moderators are oxamide, nitroguanidine, lithium fluoride *etc.*

Process aids like lecithin/silicone oil are used to improve the wetting of the oxidizers by liquid materials during processing. They help the manufacture of the propellant by decreasing the viscosity of the propellant slurry. Bonding agents are used to improve the bonding between the binder and the oxidizer, resulting in improved mechanical properties. Bonding agents are low molecular weight compounds having functional groups that either react with the oxidizer or have comparatively higher polarity than the rest of ingredients to provide secondary polar-ion attraction between the oxidizer and the bonding agent. Generally, 0.1–0.3% of the bonding agents are used. Curing agents used for composite propellants with HTPB as binder include toluene diisocyanate (TDI), hexamethylene diisocyanate (HMDI) and isophorone diisocyanate (IPDI). Curing agents are generally used in combination with polyhydroxy compounds like trimethylol propane (TMP). Dibutyl tin dilaurate (DBTDL) or metal complexes like ferric acetylacetonate (FeAA) are used as curing catalysts. Various other additives used in CP include anti-oxidants such as phenyl betanaphthylamine (PBNA).

To restrict burning on certain selective propellant surface, certain inert materials (polymeric) with inorganic fillers are used as inhibitors. Their details are given in Chapter 5.

The CMDB class of propellants use nitramine (RDX/HMX) to produce higher energy and non-smoky exhaust products.

3.1.3   Fuel-Rich Propellant (FRP) Ingredients

Fuel-rich propellants (FRP) generally contain a higher percentage of fuel; either metallic or polymeric. Mg/Al and their alloys, B, Zn, Ti, Ni are promising metallic fuels. Beryllium would have been an ideal fuel, but is not used due to its toxicity. Among the polymeric materials HTPB and GAP can be used as fuel for FRPs. Recently, pyrolysable polymers like polycyclopentadienes are being investigated as fuels. Naphthalene/anthracene can also be used as a fuel for pressed FRPs.

3.2   Liquid Propellant Ingredients

In liquid propellant based rockets, two pressurized tanks are used to hold the oxidizer and fuel ingredients separately as shown in Figure 1.1. The physical state of the ingredients remains liquid and they are fed into the combustion chamber through valves. The oxidizer and fuel ingredients are discussed in this section.

Liquid oxygen (LOX) has been a major liquid oxidizer component. It is non-toxic, non-corrosive and does not react spontaneously with organic materials. Hydrogen peroxide (H_2O_2) has a high boiling point, high density, low viscosity and gives high performance. Storing it in a container however, needs proper precautions as its concentration, impurities present and container materials affect its stability. H_2O_2 is an attractive oxidizer. H_2O_2 is generally employed as a monopropellant or gas generator. It has high boiling point, high density and low viscosity (70% H_2O_2). 90% H_2O_2 liberates 42% of gaseous oxygen. Nitrogen tetroxide (NTO) is generally used with nitric oxide. Nitric acid has been widely used in bipropellant systems. White fuming nitric acid (WFNA) contains less than 2% water and other impurities, whereas red fuming nitric acid (RFNA) contains 5–20% NO_2 dissolved in nitric acid. RFNA and gasoline produce higher performance than nitric acid–aniline system.

The characteristics of liquid propellant oxidizers is given in Table 3.2. A wide variety of compounds can serve as liquid fuels. The properties of some representative liquid propellant fuels is given in Table 3.3.

Table 3.2   Liquid propellant oxidizers.

Oxidiser	Boiling point (°C)	Melting point (°C)	Density (g cm^{-3})
Oxygen	−183.33	−217.78	1.14
Ozone	−111.11	−251.11	3.03
Hydrogen peroxide	152.22	−1.66	1.44
Nitrogen tetroxide	21.11	−9.44	1.45
Nitric oxide	−151.11	−161.11	1.27
Nitrous oxide	−89.44	−102.22	1.23
Nitric acid	86.11	−42.22	1.50
Tetranitromethane	125.55	12.78	1.65
Fluorine	−185.56	−223.33	1.11
Chlorine trifluoride	12.22	−83.33	1.77
Bromine pentafluoride	40.56	−61.67	2.46

Table 3.3   Properties of liquid fuels.

Fuel	Density (g cm^{-3})	Boiling point (K)
Hydrogen (L)	0.07	386.66
Ammonia (L)	0.82	20.4
Hydrazine	1.01	360.6
MMH	0.87	360.6
UDMH	0.791	336
TEA	0.723	362.6
PENTABORANE	0.63	331.5
Aniline	1.03	457.5

Liquid hydrogen has a low density and a low boiling point. UDMH is a clear colorless liquid with a density of 0.791 g cm^{-3}. It is miscible with water, ethanol and petroleum fuels. It is hypergolic with HNO_3 and can be used with liquid oxygen. A number of missiles like Nike, Rascal and Venguard use UDMH as the fuel.

All oxygen deficient compounds and those containing hydrogen come under fuel. Liquid ammonia gives a high performance with LOX but has an adverse vapor pressure of 35 kg cm^{-2} at 70 °C. Hydrazine has a high freezing point and it can be hazardous.

Nitromethane and methyl nitrate are monopropellants in which both fuel and oxidizer moiety lie in same molecule. Among fuels of bipropellants hydrocarbons, amines, hydrazines and boranes are the most common and their counterpart oxidizers are nitric acid, nitrogen oxide, hydrogen peroxide, liquid oxygen and halogens.

Among cryogenic fuels liquid hydrogen is the most common and fluorine is the most effective cryogenic oxidizer. When compared to oxygen containing oxidizers like nitric acid, hydrogen peroxide, dinitrogen tetraoxide, fluorine containing oxidizers like oxygen difluoride (OF_2) and chlorine trifluoride are more energetic and reactive. OF_2 is a colorless gas with a melting point of −224 °C and boiling point −145 °C. If the performance of (OF_2) is compared with a stoichiometric mixture of O_2/F_2, it is denser (density 1.52 g cm^{-3} at −145 °C) and consequently gives better performance.

Earth storable liquids, in addition to possessing sufficiently low freezing point have high density, low vapor pressure and relatively low viscosity. However, many popular oxidizers like ClF_3, N_2O_4, H_2O_2 do not fulfill the additional criteria. So liquid propellants have been prepared as blends of unsym-dimethylhydrazine (UDMH) with hydrazine, acetonitrile or diethylenetriamine (DETA). The corrosive nature of RFNA (red fuming nitric acid 85%) and WFNA (white fuming nitric acid 97%) can be controlled by the addition of 1.0% HF or sometimes by phosphoric acid.

Hypergolic systems are conceived by the combination of various fuel and oxidizers. A better combination is obtained if the induction period is small for the initiation of such mixtures. WFNA is a major oxidizer which gives 5 ms induction period with hydrazine. Some of the composite fuel combinations evaluated globally are given below:

MAF 1	50% DETA + 40% UDMH + 10% acetonitride
MAF 3	20% UDMH + 80% DETA
MAF 4	60% UDMH + 40% DETA
Hydrodyne V	75% Hydrazine + 25% monomethyl hydrazine (MMH)

In some of the systems some insoluble component like aluminium is added to increase the combustion energy. However, it requires the formation of a uniform suspension either by forming an emulsion or by the formation of gels using silica and acetylenic black or by using a hydrophilic polymer as a gelling agent.

Small amounts of additives are added to obtain improved ignition, lower freezing point and reduced corrosion. Spontaneous ignition occurs on gasoline–RFNA system. Amine or mercaptane is added to gasoline and a strong oxidizing agent such as $KMnO_4$ is added to nitric acid. Likewise, NH_3–RFNA can be made hypergolic by the addition of an alkali metal such as lithium. The freezing point of hydrazine can be lowered by the addition of ammonium thiocyanate. Water reduces the freezing point of hydrogen peroxide. The addition of a small amount of HF to HNO_3 reduces Al corrosion and stainless steel corrosion almost to zero. Thus, specific chemicals are added to fuels and oxidizers to obtain the desired properties of propellants.

3.3 Ingredients of Hybrid Propellants

The ingredients described above for solid and liquid propellants have been essentially adopted for hybrid propulsion. For FRPs, HTPB has been used as binders. As liquid oxidizers, both cryogenic like liquid oxygen and earth storable liquids have been recommended for hybrid propulsion, hybrid systems can employ a variety of fuel oxidizer combinations irrespective of their physical state and mutual compatibility. High energy formulations containing metal hydrides or metals with storable or cryogenic oxidizers offer the best performing chemical propulsion system. The heats of combustion of various hybrid fuels is given in Table 3.4.

Table 3.4 Heats of formation (kcal mol^{-1}) of hybrid fuels.

No.	Fuel	ΔH_c
1	Benzaldehyde thiocarbonohydrazone	810.92
2	*p*-Dimethylaminobenzaldehyde thiocarbonohydrazone	1132.86
3	*p*-Chlorobenzaldehyde thiocarbonohydrazone	552.97
4	2-Furaldehyde thiocarbonohydrazone	684.48
5	Formaldehyde thiocarbonohydrazone	434.24
6	Acetonethiocarbonohydrazone	713.94
7	Cycllohexanone thiocarbohydrazone	980.22
8	Aniline formaldehyde	894.83
9	*o*-Toludine formaldehyde	887.14
10	*m*-Toludine formaldehyde	1014.12
11	*o*-Anisidine formaldehyde	900.81
12	PVC plastisol	609.57

WFNA is used as an oxidizer for No. 1–7 in Table 3.4, RFNA for 8–11 and oxygen gas for the rest. Hybrid propulsion based systems are not in use at present because of the low delivered Isp, significant sacrifice of efficiency with thrust modulation and extremely low fuel regression rates, making their design difficult for short burning duration engines.

4

Solid Rocket Propellants: Processing Technologies

4.1 Introduction

Solid propellants are preferred over liquid and hybrid propellants in view of their safety, reliability, longer storage life and lower cost of the propulsion system. As mentioned earlier, solid propellants fall in two broad categories: homogenous and heterogeneous propellants. Homogenous propellants are popularly known as double base propellants (DBP) and contain NC, NG, a stabilizer and additives, whereas heterogeneous propellants (or composite propellants (CP)) contain an oxidizer, a binder, a metallic fuel and additives. Composite modified double base (CMDB) propellants take advantage of both DBP and CP. Generally, they contain a DB matrix, an oxidizer, a metallic fuel and additives. To achieve non-smoky exhaust products, nitramine based propellants consisting of NC/NG, RDX/HMX are used. Fuel rich propellants (FRPs) have been developed for IRR and SCRAMJET applications. FRPs generally contain an oxidizer, a binder, a metallic fuel and additives.

Solid Rocket Propellants: Science and Technology Challenges
By Haridwar Singh and Himanshu Shekhar
© Haridwar Singh and Himanshu Shekhar 2017
Published by the Royal Society of Chemistry, www.rsc.org

4.2   The Manufacture of Double Base Propellants (DBP)

Double base propellants are processed by two routes. The oldest method of processing DBP is by extrusion leading to the extruded double base (EDB) class of propellants. However, size, flaw detection methodology, safety involved and the requirement of higher capacity extrusion presses are their limitations. To overcome these limitations, casting techniques are resorted to and such propellants are called cast double base (CDB) propellants, where size and geometry are no longer limitations.

4.2.1   Extruded Double Base (EDB) Propellants

Both solvent and solvent-less extrusion processes can be adopted. For large web size rocket propellants, the solvent-less method of manufacture is adopted. However, propellants manufactured by this route are brittle in nature *i.e.* having high strength and low elongation. The extrusion process needs attention and care and remote controlled operation is preferred for propellant extrusion. The EDB propellant processing steps are explained in Figure 4.1. The various steps involved are explained below.

4.2.1.1   Mixing

Nitrocellulose (NC) of a desired nitrogen content (12.0–12.6%) is soaked in water making a slurry. NG is then added to the NC slurry and

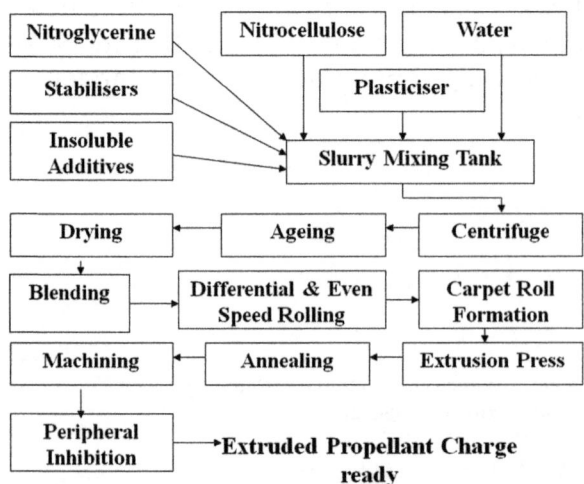

Figure 4.1   A flow diagram for the manufacture of EDB propellants.

agitated vigorously. In this process NG is absorbed in the NC fibrous matrix. Water is removed by centrifuging the slurry until it is reduced to a moisture level of 30%. The resulting NC–NG paste matured for a minimum of three weeks at 40 °C to allow the maximum absorption of NG by the NC fibers.

4.2.1.2 Incorporation of Additives

The matured wet paste after dewatering is then kneaded in a sigma blade incorporator along with additives such as a stabilizer, a plasticizer, burn rate modifiers, carbon black, a process aid *etc.* and mixed thoroughly to yield a homogeneous paste. This operation is generally carried out at 30 °C.

4.2.1.3 Drying

The wet paste is then dried in a steam heated oven to reduce the moisture content to around 5%.

4.2.1.4 Rolling

The paste, which is in a powdered form, is converted into a tough plastic sheet by working on a rolling mill. The rolling operation facilitates the gelatinization process. The dried propellant paste is passed through the horizontal even speed rollers heated at 70–80 °C and a number of passes are done to get a good gelatinized sheet. With the help of a cutter, discs (dia. 100 mm) are then punched out from the sheets. The process of rolling has some fire risk. As the rolling operation is required to be attended by an operator, a water drenching system is always provided, which functions at the first sign of fire.

4.2.1.5 Extrusion

Because gelatinized compositions behave as a thermoplastic material (that is softened on heating and hardens on subsequent cooling), it is convenient to press it at elevated temperature so that the desired configuration can be obtained by using a suitable die. Extrusion is carried out using a high capacity extrusion press having provision of evacuation of air and heating of the extrusion cylinder and die holder. The required die/pin assembly is fitted to the die holder. The pre-heated propellant discs are then loaded in the press cylinder maintained at 70–80 °C. Before pressing begins, the press cylinder is evacuated to remove air which might otherwise appear as inclusion in the grain.

This is required from a safety point-of-view as the adiabatic compression of the entrapped air can cause accidental fire. When the evacuation is completed, the press is set in motion and the plunger head engages the heated disc at the pre-set speed. On application of pressure, propellant flow starts and extrusion takes place through the die. The specific pressure for extrusion varies depending upon the grain geometry, flow properties of the propellant, die temperature and propellant temperature. The pressing operation is controlled from the outside.

The outer diameter of propellant is limited to about 300 mm because of equipment limitations. Double base propellant grains of diameter beyond this limit are obtained through a casting technique. The extruded grains are annealed to remove stress concentration and brought to room temperature. These are then cut into the required length. On account of the thermoplastic nature of the double base propellant, it can be recycled/reprocessed. The rejected grains can be used for processing after slicing into thin strips. It is standard practice to marry the hot rolled reprocess material with the virgin paste, which reduces propellant wastage.

4.2.2   Cast Double Base (CDB) Propellants

Processing of cast double base (CDB) propellants involves two main steps: (1) the manufacture of the casting powder and (2) a casting/curing operation. Additives (ballistics modifiers/catalyst) that are incorporated and the process parameters adopted during the casting powder manufacture control the ultimate performance of this class of propellants. The ballistic properties of CDB propellants are highly sensitive to the processing conditions employed during the manufacture of the casting powder. The effect of the method of casting powder manufacture on ballistic properties is sometimes greater than the effect of the ballistic modifier itself. Furthermore, during the casting process, the loading density of the casting powder, the casting powder to casting liquid ratio, the curing temperature and the time *etc.* can influence the physico–chemical and ballistic properties of the propellant.

4.2.2.1   *Manufacture of Casting Powder*

The important steps involved in the casting powder manufacture are dehydration of NC or NC–NG paste making, incorporation, extrusion, cutting, drying, sieving and graphiting.

4.2.2.1.1 Dehydration of NC

Water-wet NC is dehydrated by replacing water with alcohol in a dehydration press. The efficiency of dehydration depends upon the quality of the alcohol, the pressing pressure and the holding time *etc.* The moisture content of the dehydrated NC should be less than 5% and the total volatile matter should be 25–30%. The NC dehydration step is necessary for processing a single base casting powder only.

For double base and advanced casting powder processing, a NC–NG paste is used. For making a NC–NG paste, an aqueous slurry of NC is taken and to this a known quantity of NG is added and the slurry is agitated for several hours. Gradually, NG gets absorbed into NC. Water is removed from the NC–NG paste, and is dried by blowing hot air to a moisture level below 1%.

4.2.2.1.2 Incorporation

Dehydrated nitrocellulose (or NC–NG paste) is placed in a sigma blade incorporator and all the solid ingredients along with plasticizer and stabilizer are mixed thoroughly in presence of a suitable volatile solvent. The most common solvents used are mixtures of ether and alcohol or acetone and alcohol. The NC is gelatinised due to the action of solvent, plasticizer and the mechanical work of the incorporator blades. The various parameters governing the gelatinization process are:

i.   The type of solvent systems, their ratio and quantity.
ii. The addition sequence of the various additives.
iii. The temperature and duration of incorporation.

Generally, the temperature is maintained below the boiling point of the solvent at approximately 25–40 °C and the duration is kept between 2–3 h.

For the processing of the advanced casting powder, various sensitive materials like AP, RDX, HMX, *etc.*, are added during the incorporation stage. Hence, additional safety precautions should be taken at each stage.

4.2.2.1.3 Extrusion

After incorporation, the dough is pre-pressed under moderate pressure in a twin cylinder press for pre-compaction and removal of entrapped air. The pre-compacted dough is then taken to the main press ram and extruded through a multi-perforated die to obtain a bunch of propellant strands. The strands are collected and air-dried to remove excess solvent.

4.2.2.1.4 Cutting

The strands when moderately dry are taken for cutting in a guillotine cutting machine to obtain a cut length of 1 mm, approximately.

4.2.2.1.5 Drying

The cut strands called casting powder are then sieved to separate coarse and fine granules. The sieved material is then dried at 50 °C until the total volatile matter is reduced below 2%.

4.2.2.1.6 Graphiting

The dried powder is then coated with 0.05% graphite powder in a sweetie pan to impart flow characteristics to the powder and make it conductive thereby preventing the development of static charge.

The characteristics of the casting powder are one of the most important in-process checks of smooth processing of CDB propellant grains. The density of the casting powder should be as high as possible, normally higher than 97% of its theoretical density. Generally, the density achieved for propellants is around 1.56–1.6 g cm^{-3} in the case of single/double base powder and 1.66–1.75 g cm^{-3} in the case of advanced casting powders. A reduction in the density arises mostly from two sources: voids and volatiles. Voids may lead to pits and porosity in the propellant and may affect the ballistics properties. Volatiles such as moisture and solvents cannot be completely eliminated but should be reduced to a minimum level, preferably below 1.0%. The packing or screen loading density controls the ratio of casting powder to casting solvent. Thus, it directly affects the mechanical and ballistics properties of the propellant. A maximum screen loading density (SLD) is desired to achieve uniformity in the final product. To enhance SLD, the powder granules should be of uniform size and shape without any ragged or slant cut. Further, graphite coating also helps in enhancing SLD as the powder acquires free flowing characteristics. For optimum properties and the highest degree of reproducibility, all ingredients should be distributed uniformly throughout the powder granules. This is particularly true of the ballistic modifiers since the burning rate can vary dramatically, if distribution is not correct. The rate and uniformity with which casting powder granules are gelatinized by a casting liquid determine the mechanical integrity and ballistic properties of cured propellant. The gelatinizing charac-teristics depend on the extent of gelation of nitrocellulose during the casting powder manufacture and are influenced by type of solvent, quantity, time and temperature of incorporation. The gelatinization process also depends on the nature of the casting liquid, time and temperature of curing. The curing temperature and time are opti-mized to achieve maximum homogeneity and mechanical integrity.

4.2.2.2 Manufacture of the Casting Liquid

The casting liquid is a mixture of nitrate esters, *e.g.* NG, inert plasticiz-ers like diethyl phthalate, Triacetin, *etc.*, and a stabilizer like carbam-ite, 2-nitrodiphenyl amine *etc.* Inert plasticizers are used to dilute the nitrate esters so that their sensitivity to shock and friction is reduced considerably to facilitate handling. The inert plasticizer and stabilizer content in the casting liquid are maintained between 18–25% and 0.5–2.0%, respectively. The casting liquid is dried under reduced pres-sure before use.

4.2.2.3 Processing of the CDB Propellant

Processing of CDB propellant comprises of mould filling, evacua-tion, casting, curing, mould disassembly, machining, inhibition and testing.

The dried casting powder (TVM < 1.5%) is filled in a cylindrical mould consisting of a core, a base plate and a top plate *etc.*, using a special apparatus called a screen loader assembly to achieve maxi-mum loading density. The filled mould is then evacuated to remove volatiles and interstitial air. The casting liquid is also evacuated in similar way. This step eliminates the possibility of occurrence of voids and pinholes in cured products. In this step, the casting liquid is fed to the casting powder bed either from the bottom or the top. The rate of feeding of casting liquid is controlled to avoid the formation of channels in the powder bed.

The cast propellant is cured at ambient temperature for 48 h and at 60–65 °C for 96–120 h depending on the shape and size of the grain. During ambient temperature curing, the casting powder granules absorb the maximum quantity of the casting liquid necessary to make them swell. As the curing progresses, inter-diffusion of polymer and plasticizer occurs and the two component systems unite to form a homogenous product.

The mould is disassembled in special fixtures to remove the core as well as the cured propellant. The extraction process is highly hazardous and is generally carried out by remote control operation in a special extraction jig. The ends of the grains are soft as they are soaked in excess casting liquid. The ends are cut using a cutting saw made of non-sparking materials. The grain is then cut to the required length and machined to the proper shape and size using a lathe, milling, or drilling machine *etc.* All the operations are generally carried out under water to avoid heating during machining. In case of a propellant containing ammonium perchlorate the tools are cooled with chilled air as use of water will dissolve the water soluble oxidizers. The machine wastes are promptly removed from the machining bay, as these are fire hazards.

Most of the grains need to be coated externally and in some cases internally using polymeric materials like polyester, polyurethane, silicones *etc.*, capable of curing at room temperature to achieve a definite pattern of burning. These polymeric materials are called inhibitors and the process of coating is called inhibition. Inhibition can be carried out by coating, casting, injection, moulding, *etc.*

The finished propellant is inspected for dimension, weight, *etc.* Checking internal defects like cracks, voids, porosity, foreign matters, debonding of inhibitor, *etc.*, is carried out using radiography or ultrasonography techniques. Mechanical properties like ultimate tensile strength, compression strength, percent elongation *etc.*, of the propellant are also determined to estimate the propellant's integrity.

4.3   Processing of Composite Propellants (CP)

The CP manufacturing system can be introduced with the help of a typical block flow diagram as shown in Figure 4.2. Having established the satisfactory or within specification ballistics and mechanical properties of a new propellant system, a manufacturer is then faced with a reasonably limited choice of manufacturing methods.

It is possible to generalize that the best manufacturing system is the one that has the greatest yield of raw materials mixed or combined in a manner that provides the designed ballistic and mechanical performance in the finished product. The established production method, steps include hardware preparations, raw material processing, propellant mixing, casting assembly, casting of propellant, curing and post curing operations.

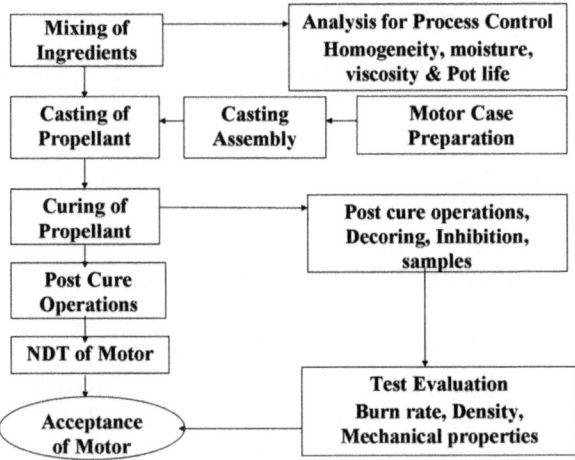

Figure 4.2 A flowchart for processing case bonded composite propellants.

4.3.1 Hardware Preparations

Hardware preparation of the motor cases is a very important step as it provides the required thermal protection and means for bonding of propellant with the motor case. Thermal protection is provided through insulation inside the motor case and the propellant's bonding with the insulators is achieved *via* abrading and liner applications.

4.3.2 Insulation Laying

The thermal insulator is an internal layer between the case and the propellant grain made of a rubber sheet, a thermally insulating material that will not burn readily. Its purpose is to limit the heat transfer from propellant combustion zone to the case during rocket operation. This is achieved by using tough elastomeric materials such as neoprene or butyl rubber, which are chemically resistant to the hot gases.

Before insulation laying, the inner surface of the motor case is roughened through grit blasting, where cast iron grits are pneumatically conveyed through a nozzle at high speed and blast the inner surfaces creating surface roughness and increased surface energy. After grit blasting, surfaces are cleaned thoroughly and a primer is applied. The calculated thickness of the insulation layer is provided on the blasted surface and the insulator is cured in an oven at an elevated temperature. Insulation laying should be carried out immediately after grit blasting to avoid depletion of surface energy and hence bonding capability.

4.3.3   Abrading and Liner Coating

The propellant does not bond with the insulator without liner as a coating adhesive. It consists of a binder and curing agents (same or similar to that of propellant), C-black or other fillers *etc.* It forms a physico–chemical link between the propellant and the insulator (physically bonded with the insulator and chemically linked with the propellant *via* covalent bonding).

To improve the bonding between the liner and the insulator, the insulator surface is abraded using a flexible shaft grinder with sandpaper wheels fitted to a flexible shaft. After abrading, the abraded rubber dusts are cleaned thoroughly by industrial vacuum cleaner and then a volatile solvent (such as trichloro-ethylene). The abraded motor is then dried in an oven at an elevated temperature to ensure the removal of the volatile solvent. The motor is then taken for liner application whereby the liner material, in the form of viscous liquid, is applied on the insulator surfaces through spraying arrangements.

4.3.4   Propellant Mixing

The next operation is propellant mixing, which is more complex and critical than the other processes in determining the quality and performance of the propellant. There are different types mixing systems used for the production of CP.

Recently, there has been an increasing use of a vertical change can, planetary mixers. Practically all of the most recent manufacturing facilities for propellant mixing that have been placed on stream have utilized large (up to 420 gallon) change cans, vertical planetary mixers of either two or three blade design. Figure 4.3 shows a typical two blade planetary mixer. In these mixers, the propellant ingredients are mixed in accordance with a pre-determined mixing cycle. Any change in mixing parameters such as sequence of addition, mixed duration, blade RPM's, vacuum conditions, charging method *etc.* may lead to undesirable rheological characteristic and final grain properties. Propellant mixing is carried out in two stages: (1) pre-mixing and (2) final mixing. In the pre-mixing stage, all the ingredients except for the curing agent are added and mixed to attain homogeneity.

Under these conditions, the propellant slurry can be kept in storage for a few days. In the final mixing stage, the curing agent is added and mixed. The propellant slurry, after final mixing, starts gelling and the viscosity starts increasing due to polymerization and cross-linking reactions.

Figure 4.3 A two-blade vertical planetary mixer.

4.3.5   Casting of Propellant

The assembly of casting fixtures is a significant feature as it provides specified grain shape, and facilitates flow of propellant slurry inside the motor case. First of all, the motor case is made vertical. The Teflon-coated mandrel or core that provides the required grain shape is assembled using a centering ring and other tooling supports. A cone and a splashguard are provided and the assembled motor is then vertically kept inside a casting chamber. The propellant slurry feeding line is then assembled within the chamber and the system is kept under vacuum for a few hours.

A typical propellant casting system is shown in Figure 4.4. The casting bowl, containing the propellant slurry, is placed on a movable trolley. It is linked to the top of the chamber by a duct. The end of this duct opens into the casting chamber, above the motor case. It is equipped with a flow control valve to control the casting rate and a slit plate. According to the design of the slit plate, the propellant slurry takes the shape of fillets or ribbons of slurry, which pile up inside the motor cases. The important factors that govern propellant casting are pot life or the castable life of the propellant, the casting rate and the chamber vacuum. Optimizing the casting rate is a trade off between (1) the need to have a rapid flow to accommodate time saving industrial requirements and to avoid significant

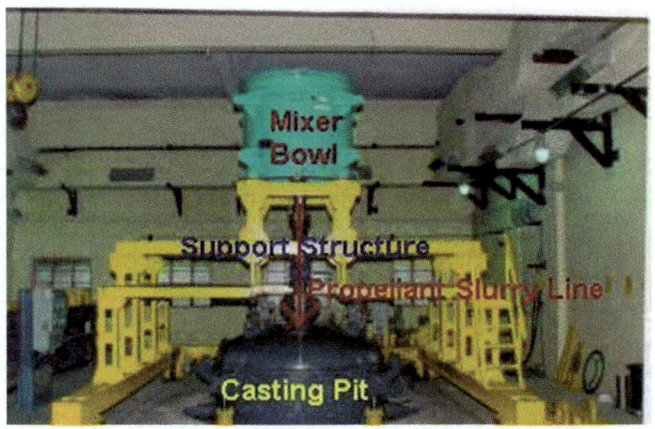

Figure 4.4   Propellant casing set-up.

viscosity build up due to progress of curing and (2) the need for a fairly slow flow to permit sufficient degassing of the slurry after it has gone through the slit plate.

4.3.6   Propellant Curing

After casting, curing is a very important operation where the propellant slurry becomes a solidified mass due to the high degree of polymerization and cross-linking. As curing progresses, the rheological characteristics, wetting properties, extent of polymerization and mechanical properties of the propellant continually go on changing until a final cure is achieved. From a practical standpoint, a system is considered to be completely cured when its mechanical and ballistic properties do not show change with respect to a moderate time scale on storage. Typically, curing is carried out at elevated temperatures during which the propellant grain, in the form of slurry, is kept inside an insulated oven for certain period of time known as the propellant curing cycle. Elevated temperature curing requires an insulated oven for monitoring and control of temperature during the curing cycle.

A Teflon-coated metallic (steel, aluminium) mandrel or core is used during casting to give the required grain shape (*e.g.* star, cylindrical *etc.*). After curing, the grains attain the required shape and the mandrel is extracted by a process known as decoring. During the assembly of the casting set-up the mandrel is properly lubricated to facilitate easy removal after curing.

Figure 4.5 An end trimming machine for propellants.

4.3.7 Propellant Trimming

It is very difficult to cast the propellant to an exact dimension because of porosity at the top of the cured solidified propellant up to certain depths and dimensional changes occurring during elevated temperature curing. Case bonded grains can be cast only once and, therefore, some extra propellant is cast to bring the grains to the exactly required dimensions. Extra propellant is trimmed off using a special purpose propellant cutting M/C, as shown in Figure 4.5.

4.3.8 Loose Flap Filling

Loose flaps are provided on either end of case-bonded composite propellant rocket motors as they provide space for dimensional changes in the grain due to thermal expansion or shrinkage during propellant curing and thus avoiding undue thermal stresses. The gaps between loose flaps and insulator rubber need to be filled up by suitable materials (epoxy *etc.*). This is done to avoid localized burning and oscillating pressure load created in gaps during burning.

4.3.9 Inhibition

Depending on the type of application, motors are used for three different burning patterns: (1) progressive, (2) neutral and (3) regressive. To provide the control over the burning pattern area, the burning surface is controlled. This is achieved using inhibition. An inhibition material is usually inert and does not readily burn along with the propellant. The choice of inhibition material depends on the type of propellant and its composition. In this operation, the material is usually prepared to form a viscous liquid and cast over the prepared propellant surface. It is cast layer-by-layer to attain the desired dimensions and geometry.

4.4 Processing of Composite Modified Double Base (CMDB) Propellants

The CMDB class of propellants are processed either by a slurry cast technique (SCT) or by an advanced casting powder (ACP) route.

4.4.1 The Slurry Cast Technique (SCT)

The slurry cast route for processing CMDB propellants has gained major importance as it enjoys flexibility of composition, ease in processing large size motors and intricate shaped complex designs. In this case, fibrous NC, which is generally used in extruded propellants, cannot be used because the plasticizer converts NC into a non-castable dough on gelatinization. The fibrous NC has to be converted to partially gelatinized spheroidal NC (SNC) before incorporation into other components to form a free flowing, non-settling slurry. This fluid slurry is deaerated and cast with ease similarly to composite propellants.

The manufacture of SNC begins with the dissolution of fibrous NC in ethyl acetate. NG, plasticizer and stabilizers are directly added and mixing is continued to get a uniform gel. This is emulsified with water in a disperser in the presence of a wetting agent and a colloidal stabilizer. Ethyl acetate (solvent) is removed by equilibrium distillation and the gel hardens slowly. For facilitating water removal, sodium sulfate solution is added before complete hardening. The temperature of the gel is raised to 70 °C and SNC balls are obtained. The process parameters are adjusted to achieve a particle size of 25–50 μm as a lower particle size leads to difficulty in loading and a higher particle size leads to a non-homogenous mix. After this, the casting liquid containing the NG with the plasticizer and stabilizer is prepared. The SNC is added to the casting liquid followed by other solid ingredients. Mixing is continued under vacuum and the mixing time and temperature is adjusted as per batch size, solid content, pot life and physical nature of the ingredients. Once a homogenous mix is obtained, the mixed material is cast in a pre-assembled mould under vacuum. The mould, along with cast propellant, is placed in oven at an elevated temperature and curing starts. At a micro level, the polymer and plasticizer dissolve into each other, leading to a rapid swelling of NC globules. At a macro level, curing involves gelation, which converts a two-phase solid liquid system to a solid propellant as a result the diffusion of the casting liquid into the SNC. As large quantity of solid ingredients is present, settling of these particles during curing is also observed. Hence, curing is accelerated. For relieving internal thermal stresses, the propellant is cooled slowly, once ballistic and mechanical properties have been achieved.

4.4.2 The Advanced Casting Powder (ACP) Route

Advanced casting Powder (ACP) route is more like CDB propellant route. It can be applied for large diameter propellant grains and also for intricate designs. The process starts with manufacture of advanced casting powder (ACP), followed by casting and curing. ACP comprises of all the NC and solid ingredients (oxidizers, stabilizers, ballistic modifier) with a portion of NG, AP and Al. ACP is obtained in 1 mm × 1 mm size by adopting solvent extrusion process. Initially NC is introduced as alcohol-wet material in the form of lumps and subjected to heavy agitation. A solution of NG in a volatile liquid (acetone) is added and premixng is done in a low shear mixer. After this pre-mixing, final mixing with all the ingredients is conducted in a sigma blade mixer. Once premixing is over, dough is obtained, which is subjected to blocking for consolidation and removal of trapped air. This is extruded in the form of strands, which are cut to required sizes. The obtained ACP is dried and glazed by graphite powder in a tumbling barrel. It is filled in mold with a packing density of 65–70%. Casting liquid is filled under vacuum from the bottom of the mold. Propellant is then allowed to mature for 4 days and molds are shifted to curing oven for solidification of propellant charges.

4.5 Processing of Extruded Composite Propellants (ECP)

Extruded composite propellants (ECP) are a relatively new entrant in the propellant-manufacturing arena. ECP manufacture seeks the development of composite propellants based on high-density elastomers like fluoropolymers (Viton, Teflon) or thermoplastic polyurethanes. The resultant propellant has a higher density (1.98 g cm^{-3}) than conventional composite propellants. This type of propellant eliminates wastage of propellant completely and can be recycled. They exhibit excellent dimensional stability and can be effectively used for bulk production of small size propellant charges of gas generators or power cartridges.

The major ingredients of ECP are the same as conventional CP except for the binder, which is an elastomer. ECP uses solvents like acetone, ethyl acetate (EA) and methyl ethyl ketone (MEK). The processing of ECP is similar to the processing of EDB propellants. It has three major steps: (1) solvation of binder and mixing, (2) rolling of propellant sheets and (3) extrusion of propellants.

The weighed quantity of binder is loaded in a planetary mixer bowl and solvent added in a ratio of $1:2$ by weight. A soaking time of about 24 h is allowed. The soaked binder is crushed and uniformly dispersed in the solvent by stirring. Additives like process aids and plasticizers are added and mixed. Subsequently, Al and 2–3 aliquots of AP are added and mixing is continued until homogeneous dispersion of metallic fuel and oxidizer in ensured in the dough. All these operations are carried out in a controlled atmosphere with relative humidity not exceeding 60%. The dough is unloaded in trays and is dried at around 60 °C for evaporation of solvent for sufficient time (4–8 h).

After solvation and mixing of the ingredients, the propellant mix is further homogenized by rolling out on a rolling mill at 55–60 °C. The propellant sheets are passed through rollers 25–50 times to ensure uniformity. The circular discs are punched out of these sheets for extrusion.

The extrusion of ECP grains is carried out in a hydraulic press. The die and pin of a suitable size are used to get strands for burning rate measurements. The propellant discs are loaded into press basket maintained at 55–60 °C and are extruded like the EDB grains by applying pressure.

4.6 Processing of FRPs for Ramjet/Scramjet Applications

FRPs are made either by pressing techniques or by a casting technique. In both techniques the oxidizer and metallic fuel preparation is the first step. Their mixing is carried out at 50 °C under vacuum. This is followed by casting or pressing under vacuum or *in-situ* pressing. The next step is the curing of the processed propellant grain at 70–80 °C, followed by finishing, QC checks, evaluation and storage.

4.6.1 The Pressing Technique

All the FRPs ingredients are verified and kept ready under sealed conditions. The mixing of the ingredients in a pre-decided sequence takes place and some solid lubricants (graphite) are also incorporated for ease of processing. The actual operation is incremental pressure moulding in a split die. A hydraulic press is used to apply pressure for achieving the desired density and compressive strength. After pressing, the propellant grain is extracted from the mould and the surface is cleaned with toluene to remove lubricant, adhering to the surface of mould. After this, the grains are inhibited with a suitable material (filled polymers).

4.6.2 The Casting Technique

This process is similar to processing the composite propellant. A solid oxidizer and a metallic fuel are ground to the required particle size and are kept ready in a sealed container after drying. The binder, after QC checks, is also kept ready. Similarly, the plasticizers, burn rate modifiers, bonding agents, curing agents, process aids and cross-linking agents are prepared and marked for processing. All ingredients are mixed in a sigma blade mixer. The mixer must have provision for hot/cold water circulation, vacuum application, bursting discs and a remote control panel. The sequence of mixing, mixing time and interval between each installment of mixing, temperature levels, vacuum application are controlled to attain the required end-of-mix viscosity of resulting slurry. Since FRPs have higher binder percentages, they have adequate viscosity to fill the mould by gravity alone. But vacuum casting is preferred to get void-free propellant grains. Casting is carried out in an evacuated assembled mould around the mandrel. A vacuum level of 2–3 mm Hg is maintained during casting to eliminate any voids or occluded gases. After casting is over, the filed moulds are transferred to ovens for curing and propellants are generally cured at an elevated temperature (60–70 °C). Curing is continued until the desired mechanical properties in the grains are achieved. After this, the propellant grains are trimmed, machined and inhibited.

5

Insulation, Liner and Inhibition Systems

5.1 Introduction

The solid propellant rocket motors used in launch vehicles, rockets and missiles have two main class of systems, one propellant that propels the whole system and the other is a mixture of non-combustible (casing, nozzle) and combustible sub-systems, which helps the propellants in performing efficiently. These combustible types of auxiliary inert sub-systems are less than 4% of the total weight of the flying object and are generally polymeric or elastomeric in nature. These substances include insulation, inhibition and the liner sub-system of propulsion motors. Composite propellants are employed in most of the currently deployed rockets and missiles. Earlier, it was customary to use a cartridge loaded type of solid propellant grains for propulsion in most of the rockets and missiles, because of design simplicity, charging, handling ease and replenishment option. In these cases, polymers are applied as inhibitors to selectively restrict the burning surface of the processed propellant grains for getting the desired ballistics in actual operations. For better utilization of chamber volume, especially for large size rockets, missiles and launch vehicle boosters, case bonded type of composite propellants are preferred. In these cases, insulation, inhibition and the liner got high importance. Insulation is generally applied as a thermal protection layer at the inner

Solid Rocket Propellants: Science and Technology Challenges
By Haridwar Singh and Himanshu Shekhar
© Haridwar Singh and Himanshu Shekhar 2017
Published by the Royal Society of Chemistry, www.rsc.org

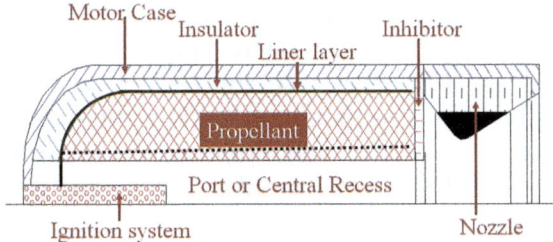

Figure 5.1 A cross-section of a typical case-bonded rocket motor.

surface of the motor casing to prevent its exposure to high temperature combustion gases (3000 K). Inhibition is generally applied at the ends of the propellants to restrict the burning surface. Liners are applied as a thin coating at the propellant–insulation interface for better adherence and prevention of migration. Figure 5.1 depicts all systems of a typical case bonded rocket motor.

5.2 The Insulation System

The main purpose of applying insulation is to withstand the high temperature combustion gases and for shielding the motor casing from heat. Depending on the propellant grain configuration, motor design and insulation properties, the thickness of insulation is decided and as a rule of thumb, it is less at the cylindrical portion and at a maximum at the end domes. To avoid the propellants cracking at the ends during curing, loose flaps are provided at the ends. They are bonded to insulation on the inner side of the motor circumferentially and at the open side, they are free to move along with propellant for nullifying shrinkage during curing. During the application stage, the elastomeric material of insulation remains in green form, which is cured *in situ* in the rocket motor using an autoclave under high pressure and high temperature. One of the major functions of the insulation is to relieve the stresses arising out of differential expansion between the case and the propellant during processing and storage.

The mechanical properties of the casing materials degrade rapidly over a threshold temperature level (400–500 °C). Nozzles are also protected by a lining of silica phenolic or carbon phenolic layer. During propellant combustion in rocket casing, a shield is essential to restrict any rise in casing temperature. For rocket motors with cartridge-loaded propellants, a ceramic coating is provided at the inner surface of the casing as a thermal shield. For case bonded applications, insulation materials are sandwiched between casing and propellants. Insulation materials are generally based on elastomers filled with inorganic fillers. They provide an interface with the rocket casing materials like 15CDV6, maraging steel, composite materials and propellant having a wide range of mechanical, thermal, chemical and ballistic properties. The selection criteria for development of a good insulator system are enlisted below.

- Bonding with propellant and motor casing should be excellent within the entire range of working temperatures and should be independent of the curing characteristics of insulation.
- It should have a low ablation rate (less than 0.2 mm s^{-1}).
- It should have a low density (less than 1.2 g cm^{-3}) for a reduced weight penalty on propulsion unit.
- It should be compatible with live constituents of the rocket motor.
- It should have low moisture absorption and good ageing characteristics.
- It should produce porous char after pyrolysis for better retention characteristics.
- It should have good strain capability to restrict de-bonding between dissimilar materials and to withstand stresses during processing, storage and handling.
- It should have enough flexibility to reduce process time and to achieve uniformity in application.
- It should have better thermal properties. Thermal conductivity should be low (less than 5×10^{-4} cal cm^{-1} K^{-1}) and specific heat should be high (more than 0.4 cal g^{-1} K^{-1}).
- There are certain auxiliary requirements also: the cost of compounds and fabrication should be low; storage and handling should be safe; design and fabrication should be as per current state-of-the-art technology.

The insulation recipe consists of an organic rubber with suitable fillers. The organic rubber can be nitrile, ethylene propylene diene monomer (EPDM), natural rubber, polyisoprene *etc.* The fillers are generally precipitated silica, carbon black, calcium carbonate, asbestos, mica powder, clay *etc.* Several other additional additives are also incorporated in typical insulation formulations for imparting specific properties *e.g.* activators (zinc oxide, stearic acid), curators (sulfur, peroxide), tackifiers (coumorine resin, terpene-phenol resin or capolyte) for bonding improvement, vulcanization activator (mercapto benzo thiozole), flame retardants (antimony trioxide, tricalcium phosphate), process aids for improving flexibility (paraffin oil or plasticizers) *etc.* Sulfur-vulcanized insulation are incompatible with CTPB based propellants. Therefore, peroxide based insulations based on BUNA-N rubber reinforced with polyaramide fibers are used. The use of asbestos in compositions is banned because it causes the disease asbestosis. A recent recipe replaces organic rubbers with inorganic elastomers like poly dimethyl siloxane (PDMS) reinforced with short fibers of polyaramids. EPDM based elastomers, cured by peroxide and reinforced with Kevlar or carbon fiber have been used in recent years.

The insulation thickness is calculated on the basis of heat flux, time of exposure to high temperature gases, ablation rate and other thermal properties of the insulation. For different requirements, the laying techniques of insulations are different. The most common method is preparation of insulation sheets of 1–3 mm thicknesses and bond them at the surface of motor in an unvulcanised state. Additional layers can be applied, if the thickness requirement is higher. Making sheets of higher thicknesses is not preferred because of quality control related difficulties. These sheets are cured/vulcanized at higher temperatures under pressure in an autoclave for good bonding with rocket casing materials. Another method is by casting the insulation slurry around a mandrel of a suitable dimension placed in the motor. This can be accomplished either under vacuum or feeding by pressure. Adequate physical properties are developed during curing. Trowelable and sprayable type of insulation materials varies in the method of their application on the motor surface. If thick semi solid insulation is applied, as is used by a mason during plastering or used in sand casting operation, trowelable type of insulation is required. This should have adequate rheological properties to wet the surface, to spread over it and to resist subsequent deformations leading to unwanted thickness variations. Sprayable insulations are dissolved in a fast evaporating solvent carrier, which evaporates after spraying. The thickness is built-up by repeatedly spraying with sufficient dwell time in between for evaporation of the solvent.

In general, at the interface of propellant and insulation, the liner is sprayed for better bonding, but a new class of insulation material can bond directly with the propellants with the help of their own reactive groups. This is referred as insuliner or reactive insulation and a typical composition can be a silica filled HTPB rubber based material.

Although the determination of insulation thickness is difficult because of several factors like heat flux not remaining constant during motor operation; non-uniform heat transfer due to differential gas velocities; the onset of char formation and heat transfer thereafter; and erosion of insulation by hot high velocity gases *etc.* Based on factors like ablation rate, erosion rate and charring, the insulation thickness is generally decided. However, after observing the casing skin temperature and thickness of unburnt insulation in actual motor operation, the thickness is finalized. Among mechanical properties of the interface, tensile bond strength, wheel peel strength and shear bond strength with propellant and casing become major governing parameters. High values of these parameters along with high percent elongation and tensile strength have been the most desirable feature of insulations. An effective insulator for a Solid Rocket Motor (SRM) should have a tensile strength of 40 kg cm^{-2}, an elongation of 400%, a tensile bond strength of 5 kg cm^{-2} and a peel strength of about 1 kg cm^{-1}.

5.3 The Liner System

The liner is a thin layer of elastomeric polymer applied at the propellant–insulation interface to improve the bond strength between them. Although liners add extra safety in actual motor operation, their use has been very limited and often their thickness has been restricted to a maximum of 200 µm. Bond failure may occur due to cumulative effect of thermal stresses during propellant cure and stresses induced during transportation, handling and storage. In general, binders used in solid propellants are much weaker than the polymers used in liners and an adhesive failure is an undesirable event. A cohesive failure in the propellant has been always advocated. Wettability, adhesive set and the deformity of adhesive to relieve internal stresses have been set as governing parameters.

As far as bonding mechanisms are concerned, one theory highlights the polarity of liner and propellant materials. It indicates that polar adhesives will not form good bonds to non-polar adherents but this rule has exceptions. Another theory is based on electric charge transfer between sticking phases and ionic type bonds are established between adherents, having opposite charges. Sometimes actual chemical bond formation has been observed. Whatever may be the mechanism, the wettability of the substrate by the propellant is assumed to be a prerequisite for bonding and the presence of active hydrogen has been an adhesion reducer in some systems based on amines and carboxylic acids.

The major liner composition requirements are listed below.

- The liner is compatible with the propellant.
- The liner composition should be room temperature cured.
- The liner should have good flow properties and ageing characteristics.
- The liner must prevent inter-diffusion and plasticizer migration and should act as a diffusion barrier at the propellant–insulation interface.
- The liner system must be highly reliable, considering the small thickness.

The liner system contains a binder filled with fillers, plasticizers and curatives. HTPB has been used as a base polymeric binder along with cross-linking agents like pyrogallol or trimethalol propane (TMP). The system may be TDI or IPDI cured similarly to conventional composite propellants with the exception that the NCO/OH ratio is kept very high (1.3) as compared to propellant formulations (0.8–1.0). The use of MAPO (tri methyl aziridinyl phosphine oxide), aminopropyl triethoxy silane, triethoxy (3-isocyanato propyl) silane as an adhesion promoter have also been reported similarly to bonding agents used in propellants for the improvement of powder–binder interface properties. A typical liner formulation for composite propellants is based on HTPB, carbon black and antimony oxide with a reported density of 1.0–1.2 g cm^{-3}, hardness of 50–55 on the Shore A scale, a tensile strength of 15–20 kg cm^{-2}, elongation of 40–50% and a moisture content less than 0.5%. The interface properties reported are a tensile bond strength of more than 5 kg cm^{-2}, a peel strength of more than 0.5 kg cm^{-1} and shear bond strength of more than 5 kg cm^{-2}.

Recently, HEMRL, (DRDO-Pune) has developed and patented a HTPB based insulation system for case bonded HTPB based composite propellant motors. The major advantage of this system include, the same (HTPB) matrix for both propellant and insulator, resulting in a stronger bond and no incompatibility problems and excellent low temperature properties (glass transition temperature, $T_g = -75\ °C$). EPDM based insulation systems, which are being developed, have distinct advantages of low specific gravity, better thermal characteristics, superior low temperature flexibility and superior chemical and weathering properties. EPDM based insulation systems have been used for rocket motors of ARIANE rockets, NASA space shuttles and tactical missiles.

As far as application technology of the liner is concerned, the liner composition is diluted by a volatile solvent like di-chloromethane and is sprayed at the pre-heated insulation. The solvent evaporates leaving a thin layer of liner at the insulation surface. Sometimes, electrostatic forces are utilized for dispersion of liner at the insulation. After liner application, the rocket casing is rotated in a horizontal position to achieve a homogenous distribution of the liner. The liner remains in a tacky state during propellant casting and cures along with the propellant to form a good bond with the propellant.

5.4   The Inhibition System

Propellant grains for rockets and missiles have to meet process ability, ballistic and compositional requirements. Depending on the requirement, different types of burn area variations are sought with respect to web consumption like progressive, neutral or regressive. Inhibition is selectively applied to restrict the propellant from burning in order to get desired burn area variations, *e.g.* a tubular grain with ends inhibited can produce a neutral profile, which without any inhibition will produce a regressive profile and with inhibition at lateral and end surfaces shall produce a progressive profile (Figure 5.2). To fulfill the requirements of preset missions, inhibition helps in achieving different thrust time variation profiles with the same propellant geometry. In some cases, a thin layer of inhibition is applied at the head end of the propellant port to prevent ignition peak due to head end propellant erosion. Generally, different inhibition systems are developed for different classes of propellants. Overall expectations from a good inhibition material are as follows.

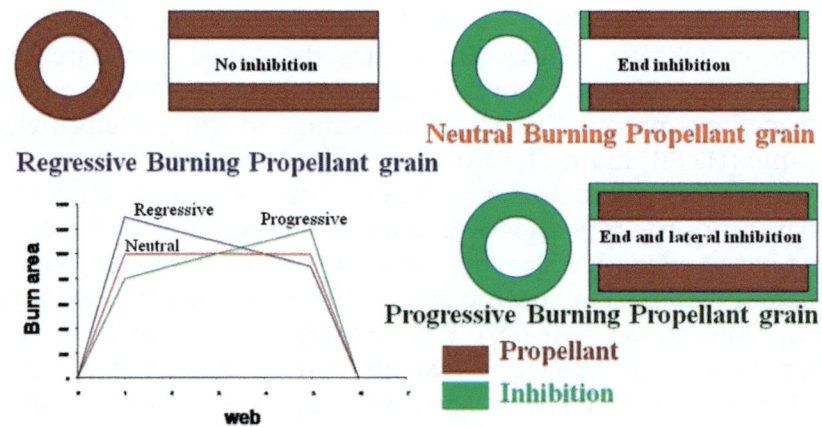

Figure 5.2   Different burning patterns using inhibitors.

- Compatibility with the propellant.
- Low density of inert mass for reduced payload.
- Low erosion rate (less than 0.1 mm s^{-1}) for reduced thickness.
- Low thermal conductivity and high specific heat for minimum heat transfer rate.
- High decomposition temperature.
- Low gel temperature and sufficient gel time for ease of application.
- Room temperature curing compositions.
- Good char retention characteristics for withstanding high velocity flow of gases without exposing propellant surface to combustion.
- Excellent bonding with the propellant.
- Good mechanical properties (tensile strength and elongation).
- Good interface properties (tensile bond strength, peel strength and shear bond strength).
- Low absorption of plasticizer and moisture.
- Good ageing characteristics.
- Long storage life.
- It should be transparent for easy detection of flaws during processing.

For smaller charges, inhibition is injected over the propellant surface, where it gels at room temperature. For smaller charges (composite propellant based gas generator) the casting of inhibition in the gap around the propellant grain placed in a mold is adopted. Flexible Epoxy resin (EP-4) with an accelerator (cobalt nephthanate) and a catalyst (peroxide) have been used for such an application. Tape winding and thread winding techniques have also been utilized for applying inhibition over a propellant surface. In tape winding, inhibition tapes are prepared from cellulose acetate, ethyl cellulose or high-density polyethylene for application over the propellant. In thread winding, epoxy resin is reinforced by nylon or rayon threads to impart impact strength to the resulting composition.

Composite propellant grains used for various Indian missiles like Pechora (SAM), Pinaka (SS) and Trishul (Multi-role) have been inhibited by this technique. However, this method cannot be used for end inhibition. Another method of inhibition is sleeve casting technology, where inhibition sleeves of adequate dimensions are prepared and slipped over the propellant with a certain primer material in between for an improvement of the bonding characteristics.

For double base propellants (DBP) inhibition, the inhibitor composition should have a high modulus. In addition, the inherent problem of nitroglycerin migration towards the propellant–inhibition interface should be prevented. Over a period of time, this may result in inhibition becoming energetic and the interface deteriorating in strength. For several strategic applications, non-smoky propellants were developed and matching non-smoky inhibition systems have been demanded. Ethyl cellulose (EC) has been used for extruded charges but EC absorbs 14–17% NG/Plasticizer at 50 °C. Furthermore, EC cannot be used for end propellants due to its low softening temperature. Polyesters are also reported to be used as inhibitors for DBP. A polyester resin containing TiO_2 as the filler has been claimed to be a good inhibitor for DBP.

For inhibition of composite propellants, bisphenol–epichlorohydrin based epoxy resins with a polyamide hardener, impregnated with a rayon thread have been used extensively. For large size case bonded motors, a castor oil based inhibition system has been established. For CMDB class of propellants, inhibition composition contains epoxy resin, butadiene–acrylic acid copolymer and inorganic metal salt. Another class of inhibitor formulation for CMDB application are based on unvulcanised rubbers (butadiene–styrene, acylonitrile–butadiene blended with PVC or ethylene–propylene copolymer) filled with asbestos fibers, powdered silica, or carbon black. Flexible unsaturated polyester based on polyethylene glycol, phthalic acid, maleic anhydride with Desmodur R as a base layer have been developed for double base and CMDB propellants.

For nitramine based CMDB propellant grains for 3rd generation anti-tank missiles of Indian origin 'NAG', a HTPB based polyurethane inhibition system has been used. This system has negligible NG migration, excellent adhesion, comparable mechanical properties, low density and has proved itself in extreme temperatures. Smokeless inhibitors based on polyethyer polyol cured with IPDI and melamine as the filler have been developed for nitramine based smokeless propellants for anti-tank missiles.

These inert systems have been an essential component of a rocket. The properties to be considered during research in this area include reduction in density, better thermal properties, better mechanical properties, especially interface properties and high compatibility with the propellant. This is an open field for research and will drive scientists to explore new and alternative insulation, liner and inhibition systems.

6

The Essence of Solid Rocket Propulsion

6.1 Introduction

A rocket is accelerated forward by reaction forces, generated by the flow of combustion gases with supersonic velocity in a rearward direction. Any rocket consists of three parts: a combustion chamber, a throat and a nozzle, as shown in Figure 6.1. The combustion of the solid propellant takes place in the combustion chamber and produces hot combustion gases. The combustion products generate pressure (P_c) in the chamber. At the exit plane of nozzle, after passing through the throat, the hot combustion products gain exhaust velocity (v_e) at the cost of pressure reduction (P_e). The outwardly flowing gases generate rearward thrust. Since the outside pressure just after nozzle exit plane is variable, depending on the altitude of operation, a pressure imbalance exists at the exit plane. Therefore, thrust is generated by momentum exchange between the exhaust and the motor and by a pressure imbalance at the nozzle exit plane. The first part of the thrust, which is generated due to momentum exchange, is mathematically expressed as $F_m = \dot{\omega}V_e$, where $\dot{\omega}$ is mass flow rate of the propellant and V_e is the exit plane velocity.

The exit plane pressure (P_e) is purely a function of the outgoing gas property (ratio of specific heat) and nozzle expansion ratio (A_e/A_t). Since atmospheric pressure is dependent upon altitude and it reduces by 6.5 kg cm^{-2} for every 1 km in height, it is impossible to adjust the

Solid Rocket Propellants: Science and Technology Challenges
By Haridwar Singh and Himanshu Shekhar
© Haridwar Singh and Himanshu Shekhar 2017
Published by the Royal Society of Chemistry, www.rsc.org

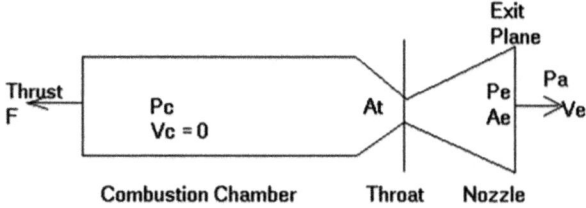

Figure 6.1   The ballistic parts of a rocket motor.

exhaust pressure level to the outgoing atmospheric pressure. If the nozzle is sending gases to the atmosphere characterized by P_a, the pressure imbalance thrust is given by:

$$F_p = (P_e - P_a) A_e$$
$$\text{Total thrust, } F = F_m + F_p = \dot{\omega} V_e + (P_e - P_a) A_e$$

If the rocket is operating in a vacuum, P_a is zero and the maximum value of thrust is obtained. The pressure and velocity of the combustion gases are interdependent and the pressure in the combustion chamber gets converted to velocity. In general, a higher value of exit plane velocity is desirable. However, the quest for higher exit plane velocity is restricted by the corresponding reduction in exit plane pressure. For a very high exit plane velocity, the exit plane pressure may attain a value lower than atmospheric pressure and thrust, due to the pressure imbalance, may become negative. This situation reduces total thrust. It is always desirable but difficult to attain a situation to get zero pressure imbalance thrust. The variation in atmospheric pressure with altitude requires different nozzle expansion ratios (A_e/A_t), which is practically impossible to obtain. However, the concept of an extendable nozzle is helpful to some extent in overcoming any deviations from an optimum thrust situation. In this case, a sliding nozzle divergent increases the exit area of the nozzle at higher altitude situations to give reduced pressure and enhanced velocity.

The specific impulse (Isp) is another important derived factor, which is an overall performance index of propulsion systems. It is defined as the amount of thrust generated per unit mass flow rate of the propellant and is a function of the specific heat ratio, the chamber temperature, the mean molecular weight of the exhaust products and the pressure ratio across nozzle. For high values of Isp, a low specific heat ratio, a high chamber temperature, low mean molecular weight of exhaust products and a high pressure ratio across nozzle are desirable. It consists of two distinct parameters: (1) characteristic velocity (C^*) and (2) thrust coefficient (C_F). Characteristic velocity represents the efficiency of the propellant combustion and is independent of the nozzle characteristics. It is a figure-of-merit for comparing propellant combinations and combustion chamber design. It is dependent on the specific heat ratio of the combustion gases and the gas temperature in the combustion chamber. The thrust coefficient (C_F) is a function of gas property, nozzle area ratio and the pressure ratio across the nozzle. It is independent of chamber temperature. A peak value of the thrust coefficient is obtained when an optimum expansion condition, depicted by the exit plane pressure being equal to atmospheric pressure, is achieved.

Total impulse is derived as Isp multiplied by the total mass of the propellant burnt. For assessing the range, payload and control environments, this is treated as a governing factor. Alternatively, this is also defined as a time integral of thrust over propulsion duration of rockets. For constant thrust and negligible start and stop, transient total impulse is defined as the product of thrust as well as time of operation.

Since the main aim of propulsion is the motion of a payload from source to target by consuming chemical energy from propellant combustion, it is mandatory to keep a watch on energy conversion phenomena and associated losses at various steps. After combustion, pressurized hot gases expand in the nozzle gaining velocity, which is instrumental in propelling the rockets. Efficiency in each conversion should be optimized to achieve a maximum value of performance parameters. Combustion efficiency is the ratio of actual and ideal heat of combustion and is generally very high (98–99%). Thermal efficiency (94–98%) is obtained as the ratio of energy available in chemical form in the propellants and the energy released to heat the combustion gases to the required pressure and temperature. Thermal insulation is provided to increase this part of efficiency. Next is nozzle efficiency, which judges the conversion of high pressure hot gases to the kinetic energy of the outgoing gases. The shaping of the nozzle, especially the throat contours (divergent shapes) is important in achieving high nozzle efficiency. The propulsive efficiency is the ratio of the used kinetic energy of the outgoing jet to the total kinetic energy of the high velocity outgoing gases. The kinetic energy available with the outgoing combustion gases is a loss and it may reach a value as high as 50% in some cases. The maximum value of propulsive efficiency is obtained when the rocket velocity is exactly equal to the exhaust velocity of outgoing gases. The performance parameters cannot be judged only by the efficiency factors and in general, a high value of thrust or total impulse requires the ejection of more mass during exhaust. This leads to an effective utilization of the limited rocket propellants available.

6.2   Thermo-Chemistry

Exothermic self-sustained chemical reactions are the basis of propellant combustion. The heat generated by initiation of the propellant is consumed in heating the non-initiated part of the propellant to a self-sustained reaction temperature. It is also utilized in heating the created exhaust combustion products of the already consumed propellant to chamber temperature. The confinement within the combustion chamber causes the pressure to rise and combustion is treated as the rapid oxidation of propellant ingredients. The major constituents of solid propellants are based on four ingredients namely carbon, hydrogen, oxygen and nitrogen. During combustion, a higher Isp is obtained when the mean molecular weight of exhaust products is small. Since carbon gives carbon monoxide (molecular weight = 28) during partial combustion with less heat release (less rise in chamber temperature) and carbon dioxide (molecular weight = 44) during complete combustion, a dilemma exists in the mechanism of attaining better performance. Similar observations are found with hydrogen, which gives water (high molecular weight = 18) after complete combustion. In general, propellant ingredients are chosen to produce fuel rich gases to reduce the mean molecular weight of exhaust gases, even at the cost of reduced energy release.

Propellant combustion is not a straight forward single step oxidation. Even one ingredient is capable of producing several products due to several parallel chain reactions taking place simultaneously. Mass conservation for each element remains valid during such reactions. Each reaction taking place during combustion has different kinetics and reaction rates. The production of each species takes place as per chemical reaction, characterized by a temperature-dependent equilibrium constant. So, the mass fraction of the generated species during combustion varies. Because of the mutual dependence of temperature and species concentration, several schemes have been proposed.

For example, the decomposition mechanism of the monopropellant nitromethane (CH_3NO_2) can produce several products, such as: CH_3, CH_4, CH_3ONO, CH_3O, CH_2O, CH_3OH, CH_2OH, HCO, CO, NO_2, HONO, NO, HNO *etc.* The associated reactions are given in Figure 6.2.

The concentration of each species varies with time and temperature. Depending on exothermic nature and equilibrium behavior of the reaction, the recombination and dissociation of product molecules can take place and the concentrations of each can vary. Table 6.1 below shows the major dissociation products of one of the major

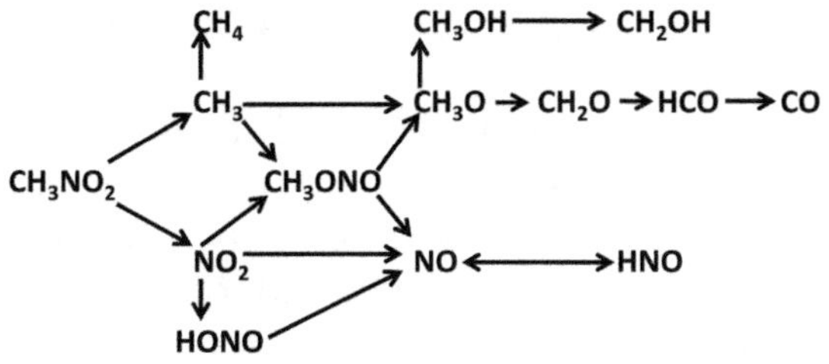

Figure 6.2 The decomposition mechanism of nitromethane.

Table 6.1 Dissociation products of NG at different temperatures.

Products	2000K	2500 K	3000K	3500K
CO_2	0.4133	0.4014	0.3381	0.2206
H_2O	0.3442	0.3380	0.3106	0.2503
N_2	0.2060	0.2029	0.1921	0.1712
O_2	0.0336	0.0355	0.0539	0.0813
OH	0.83×10^{-3}	0.59×10^{-2}	0.0231	0.0584
CO	0.36×10^{-3}	0.94×10^{-2}	0.0584	0.1460

double base propellant ingredients, nitroglycerin (NG) $(C_3H_5N_3O_9)$ at different temperatures.

Theoretically one popular approach is a Gibbs free energy minimization technique, where the initial mole fractions of probable species are assumed to a near equilibrium value and are optimized to get a minimum value of Gibbs free energy. However, tackling condensed phase species in the exhaust gases is difficult to achieve using this approach. Degeneracy exists when calculating the concentration of a condensed phase species (goes to zero) or a non-existing species is considered in the initial assumption.

For manual calculations, the initial chamber temperature is assumed in the beginning. All possible main and side reactions are considered. The equilibrium constants for each reaction at an assumed temperature are taken from the chemical reaction chart. Using this, the mass fraction of each species in the exhaust gases is determined. The cumulative heat of formation of product is obtained and the heat of reaction is calculated as the difference between the heat of reaction and the reactants. This heat of reaction is utilized to raise the temperature of the gases from the reference initial temperature to chamber temperature. If the calculated enthalpy is more than the heat of reaction, the assumed temperature is lower than the actual and a fresh calculation is done with a raised temperature until a balance between the heat of reaction and the enthalpy requirement is attained. Once the chamber temperature and species in combustion gases are obtained, the average molecular weight, specific heat, specific heat ratio and gas constant of the product gases are calculated. This helps in calculating the performance parameters in terms of Isp. Since a large number of reactions and species are to be considered, software has been developed for such calculations.

One such program, which has been utilized extensively is NASA SP-273 computer code. However, this is a purely thermodynamic calculation independent of rocket motor parameters like flow of gases, duration of combustion, and incomplete mixing that are not considered in the analysis. All gaseous components are assumed to be ideal gases and incomplete combustion aftereffects like evaporation are not considered. A typical input file for composite propellant consisting of AP(67%), Al(18%), HTPB(9%) and DOA(6%) is given in Figure 6.3.

This program calculates the concentration of all the probable species in the exhaust product of combustion at the chamber, throat and exit plane. Once the concentration of all the output species is determined, the rocket performance parameters are determined. A typical performance parameters output is shown in Figure 6.4. A typical species concentration in the combustion products of composite propellants, as obtained by the software, is given in Figure 6.5.

6.3 Nozzle Theory

The concept of increasing the velocity at the cost of pressure has been adopted from convention steam or gas turbines. Nozzles have been utilized in these systems for accelerating gases or steam to do work.

Input file for NASA SP-273

```
   1
REACTANTS
   C 7.07      H 10.65     N 0.06      O 0.22      9.00    -13800.  298.15 0.90 LF
   C 22.0      H 42.0      O 4.0       6.00                -254760. 298.15 1.10 LF
   N 1.0       H 4.0       CL 1.0      O 4.0       67.00   -70690.  298.15 1.95 SO
   AL 1.0                                          18.00       0.   298.15 2.70 SF

   NAME

   END
```

Figure 6.3   Input file for NASA SP-273.

REACTANT DENSITY= 1.7705

	CHAMBER	THROAT	EXIT
PC/P	.1000E+01	.1743E+01	.6773E+02
P MPA	.6877E+01	.3946E+01	.1015E+00
T DEG K	.3276E+04	.3065E+04	.1975E+04
RHO G/CC	.7005E-02	.4322E-02	.1750E-03
H CAL/G	-.5421E+03	-.6676E+03	-.1306E+04
S CAL/G/K	.2258E+01	.2258E+01	.2258E+01
M MOL WT	.2780E+02	.2796E+02	.2835E+02
(DLV/DLP)T	-.1010E+01	-.1007E+01	-.1000E+01
(DLV/DLT)P	.1188E+01	.1140E+01	.1008E+01
CP CAL/G K	.7254E+00	.6635E+00	.4391E+00
GAMMA (S)	.1148E+01	.1152E+01	.1193E+01
SON VEL M/S	.1060E+04	.1025E+04	.8312E+03
MACH NO.	.0000E+00	.1000E+01	.3041E+01
AE/AT		.1000E+01	.1001E+02
C* M/S		.1550E+04	.1550E+04
CF		.6611E+00	.1631E+01
IVAC KG-S/KG		.1952E+03	.2811E+03
ISP KG-S/KG		.1045E+03	.2577E+03

Figure 6.4 The ballistic parameters output from the NASA SP-273 programme.

In the nozzles, the velocity of the gases increases, while in diffusers the velocity reduces with a rise in pressure. Nozzles in rockets additionally restrict the gas flow rate thereby maintaining chamber pressure. The operating pressure of a rocket is decided by the nozzle throat diameter alone. A major concept utilized in nozzle theory is sonic velocity defined as the velocity of sound at the operating condition.

$$\text{Sonic velocity, } c = \sqrt{\gamma RT}$$

where, γ = specific heat ratio; R = gas constant; T = local temperature.

It is observed that disturbances downstream in the flow can be propagated at the velocity of sound only, and if velocity exceeds sonic velocity, upstream will not feel the variation in the downstream conditions. In general, the Mach number (M) is introduced to non-dimensionalize the flow rate. It is defined as the ratio of gas flow velocity to

AL	.6730E-04	.2324E-04	.0000E+00
ALCL	.4191E-02	.2369E-02	.1831E-05
ALCL2	.4504E-02	.3057E-02	.1465E-04
ALCL3	.2403E-03	.1933E-03	.5573E-05
ALH	.1492E-04	.4698E-05	.0000E+00
ALO	.3706E-04	.1155E-04	.0000E+00
ALOCL	.8780E-03	.5061E-03	.6929E-06
ALOH	.2724E-03	.1308E-03	.4679E-07
ALO2	.5314E-05	.1621E-05	.0000E+00
ALO2H	.2472E-03	.1175E-03	.5530E-07
AL2O	.1374E-04	.3484E-05	.0000E+00
AL2O3(S)	.0000E+00	.0000E+00	.8639E-01
AL2O3(L)	.8005E-01	.8237E-01	.0000E+00
CO	.2442E+00	.2452E+00	.2439E+00
COCL	.8547E-05	.4692E-05	.3385E-07
CO2	.9386E-02	.9339E-02	.1311E-01
CL	.5867E-02	.4592E-02	.2455E-03
CL2	.1150E-04	.7758E-05	.1445E-06
H	.2276E-01	.1694E-01	.7678E-03
HCL	.1251E+00	.1321E+00	.1474E+00
HCN	.1660E-04	.1014E-04	.3120E-06
HCO	.1692E-03	.1020E-03	.2106E-05
H2	.3285E+00	.3341E+00	.3501E+00
H2O	.9686E-01	.9318E-01	.8341E-01
NH	.6792E-05	.2918E-05	.0000E+00
NH2	.1244E-04	.6420E-05	.3339E-07
NH3	.2197E-04	.1474E-04	.1313E-05
NO	.1641E-03	.8639E-04	.2752E-06
N2	.7349E-01	.7379E-01	.7456E-01
O	.1141E-03	.5184E-04	.2818E-07
OH	.2753E-02	.1673E-02	.1739E-04
O2	.1346E-04	.5839E-05	.0000E+00

Figure 6.5 Species concentration from the NASA SP-273 programme.

sonic velocity. If flow velocity is less than sonic velocity, the flow is called subsonic. *M* equal to one indicates sonic flow and *M* more than one indicates supersonic flow.

A typical pressure variation curve in the flow through a nozzle is depicted in Figure 6.6. Initially, if the entrance pressure and exit plane pressures are equal, there is no flow exit across the nozzle. As the downstream pressure is reduced, a flow is established. For subsonic flow, the pressure reduces up to the throat at the cost of pressure but after the throat the velocity of the gases reduces in the divergent section and the pressure increases. In a subsonic situation, the divergent part of a convergent–divergent nozzle starts working as a diffuser.

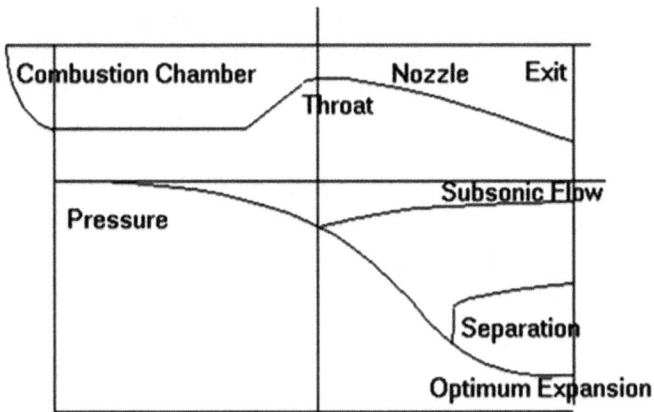

Figure 6.6 The pressure variation in a nozzle.

Sonic velocity is attained on further reduction in the downstream pressure. If flow is sonic at the throat, the upstream pressure variation curves in the convergent portion remain unchanged. This situation is called choked flow through the nozzle and the mass flow rate cannot be enhanced beyond this value by reducing the downstream pressure. Velocity continues to rise in the divergent portion and the pressure reduces. For optimum expansion, the exit plane pressure matches the atmospheric pressure and the highest performance parameters are observed. If the exit plane pressure is less than ambient pressure, the nozzle is 'over-expanded' and flow separation from the wall occurs generating shock waves and sees a significant loss in performance. The reverse situation is called 'under-expanded' nozzle, where the complete potential of combustion gases remains unutilized and performance is not optimum.

For analyzing flow through a rocket nozzle, the combustion gases are assumed to obey the perfect gas law. This assumption is good for lower operating pressures up to 100 kg cm^{-2}, but at higher pressures a correction using van der Waals' equation of state must be applied. In addition, the calculation of the equilibrium constant also needs the introduction of fugacity constants. Although during flow through a nozzle, the composition and temperature vary drastically, but for the calculation the specific heat is constant. Flow is assumed to be one-dimensional without heat loss due to friction or heat transfer to the walls. With the advent of case bonding technology, heat loss is generally confined to the nozzle section only and at the end of burning only the combustion chamber is exposed to flame or combustion gases. Propellant combustion is assumed to be completed at constant pressure in the combustion chamber only and stagnation condition depicted by zero flow velocity exists there. The process is also assumed to be steady and gases are assumed to leave the system axially with uniform velocity at any section normal to the nozzle axis.

During the analysis of the nozzle, it is assumed that a stagnation condition exists at the chamber depicted by the subscript 'c'. The throat and exit plane conditions are depicted by the subscripts 't' and 'e', respectively. Atmospheric conditions are assumed to be defined by 'a'. With these abbreviations, mass, momentum and energy conservation equations are applied.

Mass conservation equation:
$$AV\rho = \text{constant}$$

Momentum conservation equation:
$$\rho V dV/dx + dP/dx = 0$$

Energy conservation equation:
$$h + V^2/2 = \text{constant}$$

where, A = area of cross-section; V = velocity of gases; ρ = density of gases; P = pressure; x = axial distance; and h = enthalpy = $C_p T$.

The energy equation along with sonic velocity relations gives:
$$T/T_c = [1 + \{(\gamma - 1)M^2/2\}]^{-1}$$

Since $[1 + \{(\gamma - 1) M^2/2\}]$ is used frequently in the nozzle calculation, this term is expressed as 'HS' in subsequent discussions in this chapter.

Using isentropic relations, the pressure and density ratios can also be calculated.
$$T/T_c = 1/\text{HS}$$
$$P/P_c = \text{HS}^{\{-\gamma/(\gamma - 1)\}}$$
$$\rho/\rho_c = \text{HS}^{\{-1/(\gamma - 1)\}}$$

Variations in temperature (Figure 6.7), pressure (Figure 6.8) and density ratios (Figure 6.9) are plotted against Mach number for different operating specific heat ratio ranges. Similarly, the velocity ratio is plotted in Figure 6.10.

It is observed that the variation of the temperature ratio with Mach number is more dependent on the specific heat ratio (γ) and the highest change occurs for higher values of γ. However, with nozzles operating in supersonic ranges, variations in the temperature ratio with the specific heat ratio becomes insignificant. For the pressure and

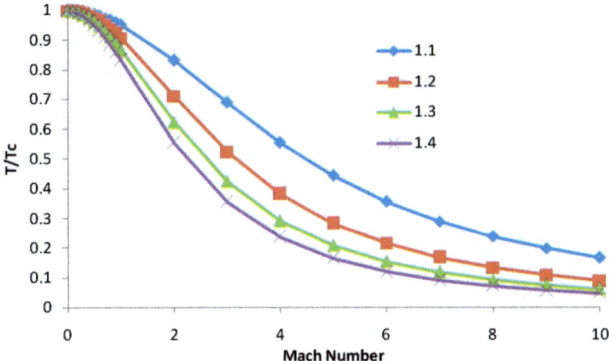

Figure 6.7   Variation of the temperature ratio with the specific heat ratio.

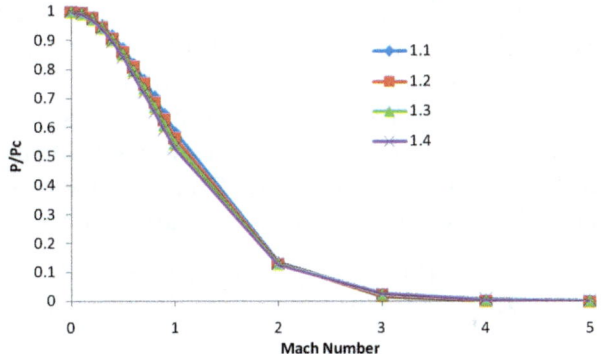

Figure 6.8   Variation of the pressure ratio with the specific heat ratio.

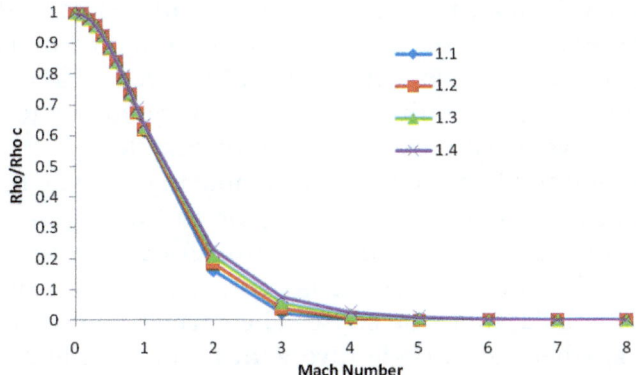

Figure 6.9   Variation of the density ratio with the specific heat ratio.

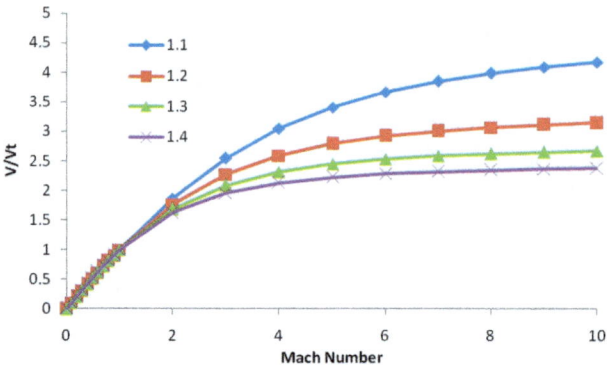

Figure 6.10   Variation of the velocity ratio during flow through the nozzle.

density ratios, variations are more or less independent of Mach number. Another important fact to observe is that the drop in pressure and density from chamber to throat is higher (50%) when compared to the drop in temperature for the same duration (10%).

In the combustion chamber, the gas velocity is assumed to be zero. So for calculating the velocity ratio, a reference point is assumed to be throat where the Mach number is one. The velocity ratio can be utilized in the law of mass conservation to get an area ratio.

$$V/V_t = M\sqrt{[\{(\gamma + 1)/2\}/HS]}$$
$$A/A_t = (1/M)\sqrt{[HS/\{(\gamma + 1)/2\}]^{\{(\gamma + 1)/(\gamma - 1)\}}}$$

When compared to the velocity at the throat or the sonic velocity, the highest velocity in a nozzle is obtained if the gases generated have a higher γ. This is derived from the fact that higher γ gives a higher temperature change and, consequently, for the same Mach number a higher rise in velocity is realized. This is also supported by the fact that the requirement of the area ratio in the divergent portion is higher for lower γ for the same Mach number rise. The area ratio curve (Figure 6.11) also justifies the requirement of the divergent portion for attaining supersonic velocity of exhaust gases during flow.

So far, no definite formulation is available to calculate the length of the nozzle and ultimately the convergent and divergent angles cannot be predicted by calculation. Deviations for calculating nozzle exit momentum of an ideal rocket needs a correction factor, since the flow of gases is not axial in a true sense. The correction factor is given by $(1 + \cos \alpha)/2$; where α = nozzle divergent half-angle. For a value of α = 15°, the exhaust velocity is reduced to 98.3%, which is applied to the momentum term of the thrust calculation. In general, a lower

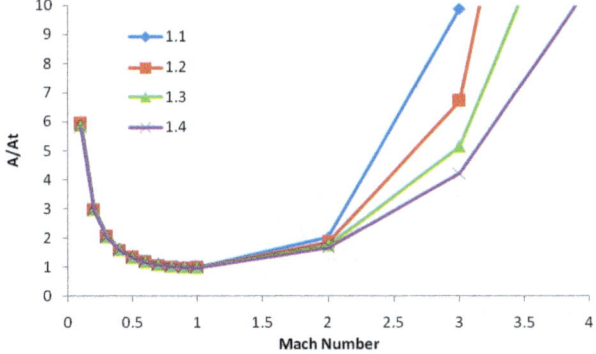

Figure 6.11 Variation of the area ratio during flow through the nozzle.

divergent half-angle results in axial flow and maximum performance (Isp) is expected but at the cost of a long nozzle. An efficient nozzle divergent should be designed as a compromise between performance (low α) and a lightweight penalty on propulsion (high α).

6.4 Internal Ballistics

Physico-chemical process inside the combustion chamber of solid propellant rockets can be explained for an ideal situation by different relationships. The key parameters include physical, chemical and mechanical properties of the propellant, combustion gas conditions, and rocket operating conditions. Three important factors, namely, the thrust coefficient (C_F), characteristic velocity (C^*) and specific impulse (Isp) are of paramount importance in rocket propulsion. To define these, the nozzle expansion ratio (ε) is conceived, which is a function of the specific heat ratio of the combustion gases (γ) and pressure ratio (P_e/P_c). Mathematical expressions are given in Figure 6.12 and variations in the nozzle expansion ratio with the pressure ratio for different specific heat ratios are shown in Figure 6.13. It is obvious that a higher value of γ gives less change in the nozzle expansion ratio for a given pressure change. The nozzle is operationally more stable and insensitive to pressure variations, if γ is high.

Thrust (F) obtained by momentum exchange and pressure imbalance at the exit plane gives an expression for the thrust coefficient defined as $C_F = F/P_c A_t$. This is independent of the chamber temperature and the mean molecular weight of the exhaust products. Except for γ (Figure 6.14), thrust is completely free from choice of propellant and depends only on the operating pressure. The optimum value of

$$\varepsilon = \frac{\sqrt{\left(\dfrac{\gamma-1}{2}\right)\left(\dfrac{2}{(\gamma+1)}\right)^{\frac{(\gamma+1)}{(\gamma-1)}}}}{\left(\dfrac{P_e}{P_c}\right)^{\frac{1}{\gamma}}\sqrt{\left\{1-\left(\dfrac{P_e}{P_c}\right)^{\frac{(\gamma-1)}{\gamma}}\right\}}}$$

$$\Gamma = \sqrt{\gamma\cdot\left(\dfrac{2}{(\gamma+1)}\right)^{\frac{(\gamma+1)}{(\gamma-1)}}}$$

$$C_F = \Gamma\sqrt{\left(\dfrac{2}{(\gamma-1)}\right)\left\{1-\left(\dfrac{P_e}{P_c}\right)^{\frac{(\gamma-1)}{\gamma}}\right\}} + \frac{(P_e-P_a)\cdot\varepsilon}{P_c}$$

$$C^* = \sqrt{\frac{g\cdot R\cdot T_c}{M}}\times\frac{1}{\Gamma}$$

$$I_{sp} = \sqrt{\frac{2\gamma}{(\gamma-1)}\frac{R\cdot T_c}{M\cdot g}\left\{1-\left(\dfrac{P_e}{P_c}\right)^{\frac{(\gamma-1)}{\gamma}}\right\}}$$

Figure 6.12   Important relationships for ballistic calculations.

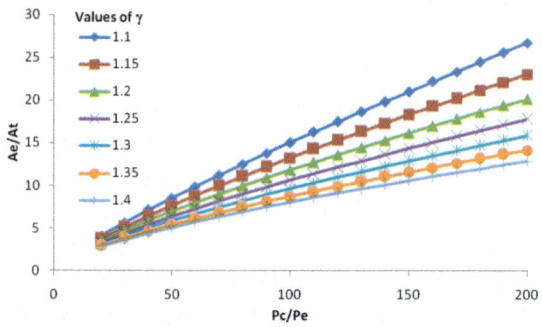

Figure 6.13   Variation of the area ration with the pressure ratio.

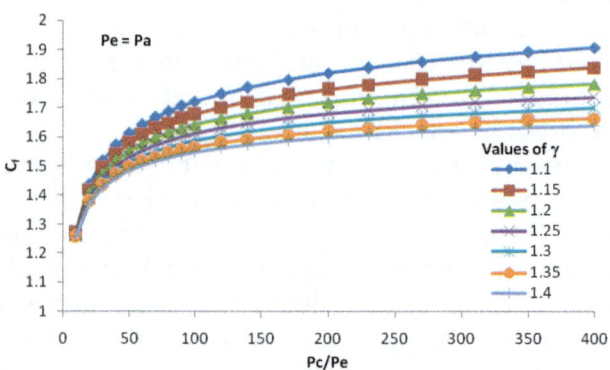

Figure 6.14   Variation of the thrust coefficient with the pressure ratio.

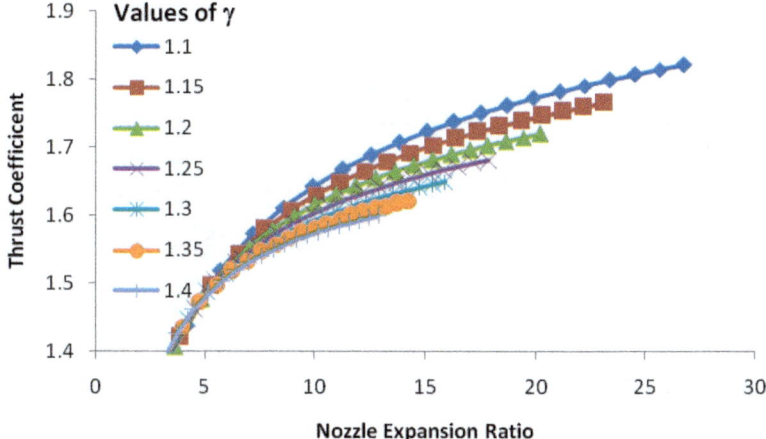

Figure 6.15 Variation of the optimum thrust coefficient with the expansion ratio.

the thrust coefficient is obtained if the rocket is operating in a vacuum and $P_a = 0$.

To get a high value of C_F, a low value of γ and a high value of nozzle expansion ratio is preferred (Figure 6.15). If thrust, due to pressure imbalance is zero, the dependence of C_F on ε vanishes. However, the trend of variation in C_F remains unaltered.

Another important ballistic parameter is C^*, which depends mainly on the conditions in the combustion chamber. It depends on T_c, M and γ, with P_c influencing it indirectly through T_c only. This parameter is sensitive to the combustion process and is a true measure of propellant performance. It is defined as $C^* = \gamma P_c A_t / \dot{\omega}$ and is independent of the downstream conditions beyond throat of the nozzle. A higher value of C^* is always desirable through a high chamber temperature and a low mean molecular weight of exhaust products. The reciprocal of C^* is called the mass flow factor and is used for the determination of the mass flow rate through a rocket.

The specific impulse is the product of C_F and C^* divided by gravitational acceleration. It is expressed in seconds and is an overall performance parameter of rocket propulsion. Similar to C^*, it needs a higher chamber temperature and a low mean molecular weight of exhaust products. It increases as the chamber pressure increases at first rapidly, then slowly (Figure 6.16).

A finite maximum value of Isp is obtained for P_e becoming zero and this value is dependent on the propellant properties, T_c, M and γ. The nozzle parameters are insignificant for this optimized value

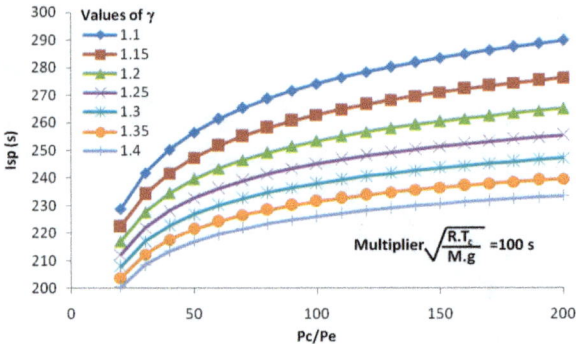

Figure 6.16   Variation of the specific impulse with the pressure ratio.

of Isp. In general, a low value of γ is desirable for a high value of Isp and when $\gamma = 1$, Isp becomes infinite. Although the value of γ for a typical propellant lies between 1.2–1.3, it does not have a large impact on Isp. If the same propellant is used at different operating pressures and the variation in γ only is considered, a higher Isp is obtained for higher pressure ratios (chamber pressure) and lower γ. The reciprocal of Isp is called the specific fuel consumption and is not very popular in rocketry. The product of Isp and density is also used as a figure-of-merit in specific cases and is called the density impulse or volumetric specific impulse. An increase in density impulse can significantly decrease the chamber weight penalty as low density propellants operate at higher chamber pressures to deliver equivalent Isp.

6.5   Rocket Behavior

Solid rocket propellant performance parameters are dependent on propellant composition, motor and nozzle design mainly. The propellant, on initiation by a suitable ignition mechanism, has to develop a self-sustained combustion reaction at the propellant surface. All propellants show a threshold lower pressure level, below which, self-sustained combustion does not take place. The ignition system should provide hot combustion gases, which should be capable of raising the temperature of the solid propellant surface to the ignition temperature and at the same time, the empty volume of the chamber (port) should be pressurized to a pressure higher than the threshold pressure for propellant combustion. For double base propellants, this limit is 30–35 kg cm^{-2}, while for a composite propellant this limit is around 15–20 kg cm^{-2}. Enough combustion gases are produced to raise the pressure to operating pressure of rockets. It is also ensured that in case of all the rocket system, choked flow at nozzle throat is realized. Once propellant surface is initiated, it burns in layers and generally follows Vielle's law of burning given by:

$$r = aP^n$$

where r = burning rate of the propellant; P = operating pressure in the combustion chamber; a = burn rate coefficient and n = burn rate index.

The operating pressure in the combustion chamber is determined by the balance of rate of mass generation by burning of the propellant and the rate of discharge the through nozzle. This balance gives an equilibrium chamber pressure:

$$P_c = [\rho a C^* A_b / A_t \gamma]^{\{1/(1-n)\}}$$

where A_b = burn area of the propellant and A_t = throat area of the nozzle.

The chamber pressure is dependent on the burn area of the propellant and instantaneous pressure values with time can be calculated for the entire combustion duration. From the chamber pressure, the thrust can be calculated directly.

$$F = C_F P_c A_t$$

From the thrust time profile, the trajectory, range, acceleration and velocity can be calculated. Of course various correction factors such as drag, lift, turning moments *etc.* must be properly accounted for before any trajectory calculations. After realizing the propellant composition, its internal configuration becomes a very important factor. From the curve shown in Figure 6.17, it is obvious that for a 90 kg cm^{-2} operating pressure, the burn area requirement is 3919 cm^2, which is possible in a solid sustainer mode for a rocket with a minimum diameter of 700 mm. For all rockets having less than such dimensions, a solid sustainer configuration is not possible to give such high operating pressure and central cavities are provided in the propellants, so that the burn area can be augmented by longitudinal surfaces.

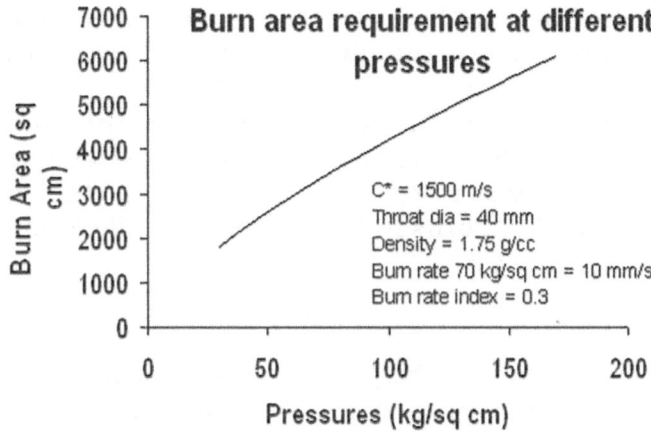

Figure 6.17   The calculated burn area requirements for different chamber pressures.

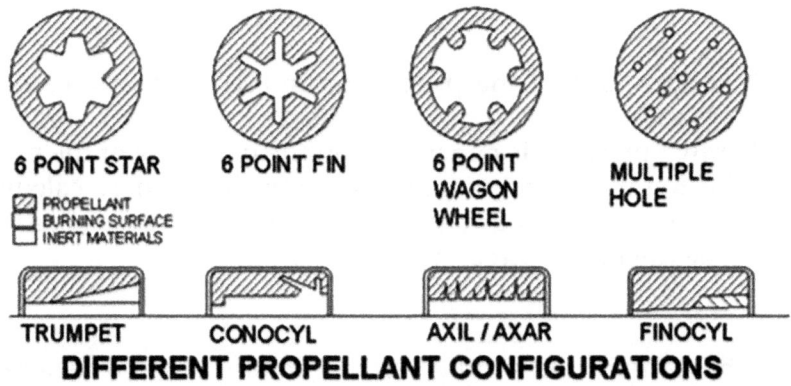

Figure 6.18   Propellant grain configurations.

Several configurations (Figure 6.18) are, in general, used in rockets such as multiple holes, star perforated, wagon wheel, dendrite *etc*. The selection is based on various requirements and specifications such as the volumetric loading fraction, web thickness, burning area, neutrality of burning, sliver fraction and web fraction. Finocyl shaped propellant grains are mostly used for higher volumetric loading fraction, high web thickness and negligible sliver fraction, where the most popular star-shaped configuration is used for moderate values of these factors. For attaining different burning patterns, bi-propellant configuration is also adopted. They are able to provide excellent neutrality and the highest volumetric loading fraction but produce lower thrust.

6.6   A Case Study

Let's consider the design of a rocket propellant for a 206 mm internal diameter motor casing of 2750 mm length. The performance parameters expected are total impulse (I_t) of 34 000 kg s^{-1} and a thrust (F) of 7500 kg.

The thrust action time of the motor can be calculated as (34 000/7500) = 4.5 s. If a composite propellant grain with a density of 1.75 g cm^{-3}, burn rate of 10 mm s^{-1}, mean molecular weight of exhaust product 22, specific heat ratio of 1.26, and flame temperature of 3200 K is considered for this situation, the web required is equal to 45 mm (action time × burn rate *i.e.* 4.5 × 10). Since characteristic velocity, C^* is a function of γ, the molecular weight and flame temperature only, it is calculated a 1666 m s^{-1}. If actual C^* is 93% of ideal C^*, the delivered C^* is equal to 1550 m s^{-1} (~1549.7 m s^{-1}). Let's consider a case bonded propellant grain with an insulation thickness of 4 mm around the circumference. This makes only (206 − 2 × 4) 198 mm available for loading the propellant. The grain design start with an assumption of the calculation of volume available for the propellant $(\pi/4D^2L)$, which is equal to 84 674.561 cm^3. If the volumetric fraction of 85% is assumed, the volume of the propellant is equal to 71 973.37 cm^3. The mass of the propellant is equal to 125 kg (~125.95 kg). If the mass fraction of the rocket motor is assumed to be 85%, the total weight of loaded rocket motor is approximately 150 kg (actual 148.18 kg). The specific impulse (Isp) needed from the propellant is equal to 272 s (total impulse/wt of propellant *i.e.* 34 000/125). Since C^* and Isp are known, the thrust coefficient C_F can be obtained as 1.724. This value is reduced by 98% to get the actual value as 1.69. This gives the value of pressure ratio of 100 and nozzle area ratio of 8.204. The motor operating pressure is 100 kg cm^{-2} for nominal atmosphere operation. The throat area can be calculated as 45.76 cm^2 $(A_t = F/C_F P_c)$ and the throat diameter is 75 mm (~7.633 cm). The exit diameter is equal to 200 mm (~20.13 cm). This completes the preliminary design operation. Next, let's look at the design of the propellant grain. A probable grain in a fin shape configuration is generated after considering several alternatives. The propellant weight is 124 kg. The pressure–time profile is also generated

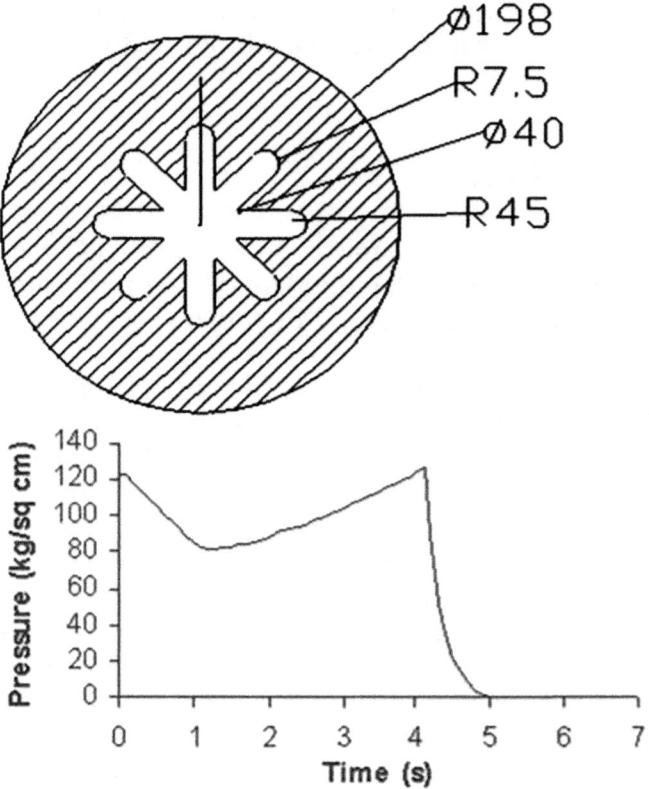

Figure 6.19   Propellant grain design: a case study.

for this grain, which meets the preliminary design specification for the propellant grain (Figure 6.19).

6.7   Conclusion

The major parameters in the selection of propellant configuration include the neutrality of the thrust–time profile, sliver fraction, volumetric loading in the combustion chamber, tail off, burn rate augmentation, and burn area with respect to throat area *etc*. In addition, for the successful performance of rockets, all sub-systems like igniter, propellant, casing, insulation *etc*. must work satisfactorily. A complete understanding and performance prediction methodology is still perfected through trials as theoretical performance predictions are based on several assumptions. All calculations are generally made with an applicability of the ideal gas law and temperature and pressure independent performance parameters. While predicting and assessing the results, it is mandatory to keep in mind the various assumptions. The generation of abnormalities, instabilities, peaks, hump effects, interface anomalies *etc.* have not been understood completely yet. However, preliminary designs of rocket propellant grains for various mission requirements can be generated easily.

7

Quality Control, Assurance and Reliability

7.1 Introduction

Quality is defined by customer needs and expectations. It is referred in several ways such as "fit for use", "fit for purpose", "customer satisfaction" or "conformance to requirements" *etc.* Since it is a vast topic, the definition can be expressed as the "Totality of features and characteristics of product or service, that bear on its ability to satisfy stated or implied needs". The 'quality' comes from a process. Any process is the transformation of a set of inputs, which can include actions, methods and operations into a desired output in the form of products, information, services or results. This is represented in Figure 7.1. To produce an output, which meets the customer/user requirements, it is essential to define, monitor and control the input as well as the process.

In the area of solid propellant production, the desired operational characteristics of a propellant can be ascertained only if the quality control (QC) is ensured for all ingredients used and at each stages of manufacturing. The various materials used in solid propellants include oxidizers, binders, plasticizers, stabilizers, process aids and ballistic modifiers. The QC of raw materials is generally done by either chemical wet analysis methods or by instrumental methods. The major QC processes are described in the next section.

Solid Rocket Propellants: Science and Technology Challenges
By Haridwar Singh and Himanshu Shekhar
© Haridwar Singh and Himanshu Shekhar 2017
Published by the Royal Society of Chemistry, www.rsc.org

7.2 QC Equipment

QC during solid propellant processing needs several sophisticated pieces of equipment. This section gives a brief outline of the different equipment used in the QC of ingredients and in-process finished products.

7.2.1 Particle Size Analysis

For particle size analysis, sieving is used to segregate the solid powdered ingredients in different particle size bands. However, to obtain particle size distribution, there several pieces of equipment available. The Malvern particle size analyzer (Figure 7.2) is a laser ray controlled device, which measures the obscuration of the ray by the particles in the ingredients. The powdered ingredients are dispersed/suspended in a suitably chosen liquid and then placed in the path of the laser.

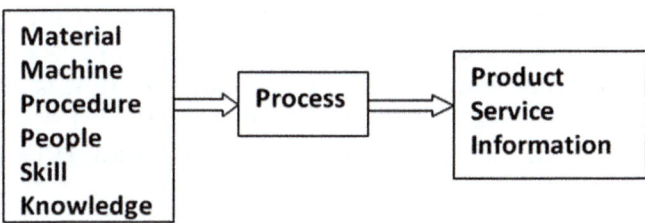

Figure 7.1 Production sequences.

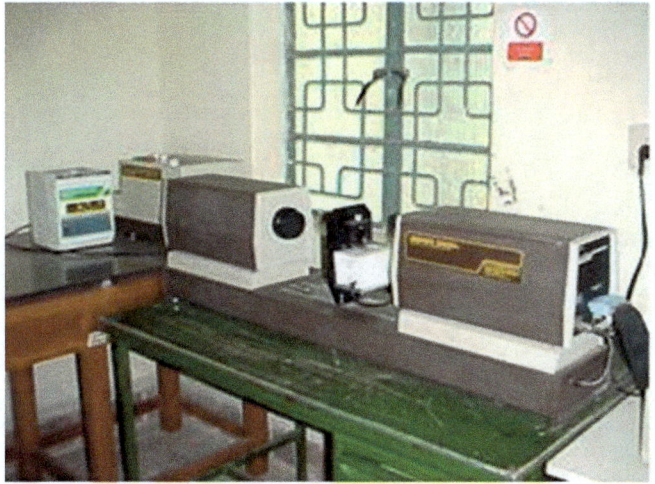

Figure 7.2 A Malvern particle size analyzer.

The time of obscuration is observed by a detector lens and is reflected as a particle in liquid (Pil) form. A sub-sieve sizer is another piece of equipment used for particle size measurement. It works on an air permeability principle for the particle size measurement of powders. Coarse particle sizes offer less resistance to the regulated flow of air for the same bed, apparent volume and percentage void. The flow of air after passing through a bed of particles is measured accurately by a precision manometer to know the particle size from a pre-calibrated chart. This is used to determine the particle size of coarse and fine oxidizer and explosives.

7.2.2 Moisture Content Measurement

The moisture content of the ingredients is generally measured using the Karl Fischer (KF) method (titration apparatus) (Figure 7.3). KF is a commercially available single solution standardized by using water in methanol solution. It has a defined measurable water equivalent (E). A weighed quantity of sample (W) is extracted with dry methanol. It is titrated with a moderate excess of KF reagent and back titrated with the water in methanol solution. The moisture content is determined by quantities of KF reagent used in titration (A) and blank run (S), amount of water in methanol used for back titration (R) and ratio of KF reagent per water in methanol solution (B).

$$\%\text{moisture} = [0.1E(A - S) - RB]/W$$

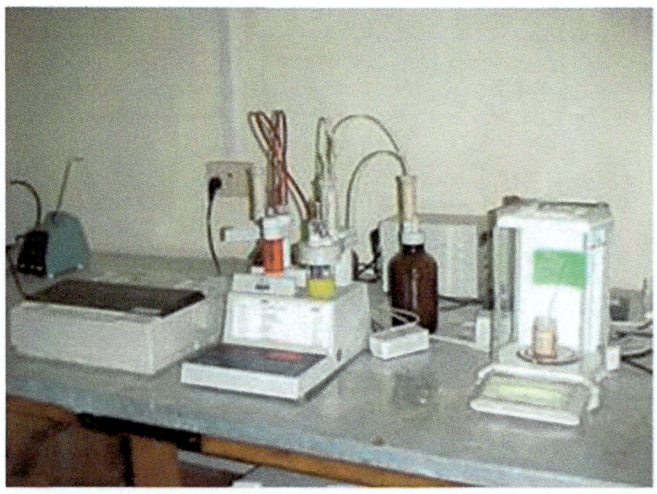

Figure 7.3 Moisture content measurement apparatus.

7.2.3   Volatile Matter Measurement

Volatile matter is measured using weighing bottles. The bottle is cleaned with solvent, dried at 110 °C and weighed. The weighed quantity of sample is placed in open bottles at 105 °C for 3 h. The bottle is removed from the oven, stoppered and placed in a desiccator for 30 min. The difference in weight yields the weight of volatile matter. This is a major QC parameter for NC, double base propellants and binders of composite propellants.

7.2.4   Viscosity Measurement

Measuring the propellant slurry viscosity is another major QC check for assessing the end-of-mix viscosity as well as viscosity build-up. Viscosity measurements need a viscometer, where a rotating spindle is immersed to the specified mark in the propellant slurry. The resistance to rotation is calibrated to give a viscosity of mix directly. The immersion of the spindle, type of spindle, rotating speed and temperature are the controlling parameters during this operation. Binder-cum-fuel (HTPB/CTPB) viscosity is linked with their molecular weight and molecular weight distribution. A Brookfield viscometer (Figure 7.4) is used for measuring the viscosity of the binder as well as the propellant slurry as an in-process QC check.

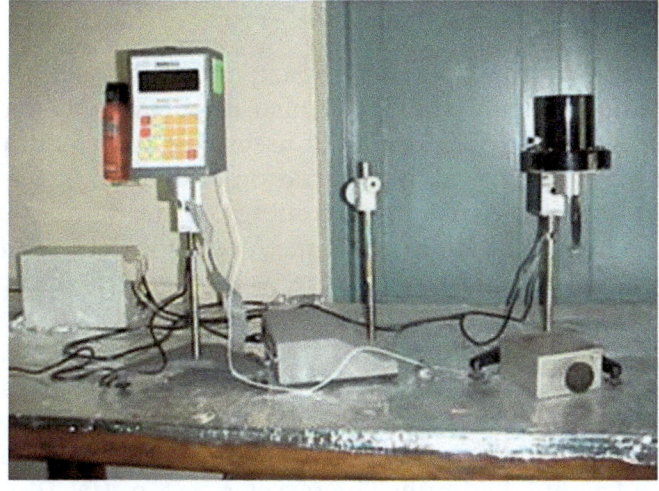

Figure 7.4   A Brookfield viscometer.

7.2.5   Density Measurement

Density is measured using the Archimedes principle. A properly cut and surface coated sample of known volume and weight is inserted in water and the reduction in weight gives the density of the sample. However, correction factors need to be applied for the coating material.

7.2.6   Calorimetric Value

A known weight (W) of the propellant sample is ignited in a semi-automatic adiabatic Julius Peter bomb calorimeter. The heat liberated is measured by a rise in temperature of water in the jacket of known water equivalent. It is mathematically equal to (water equivalent × rise in temperature − correction/W), where corrections are applied for the igniter wire and occluded air in the bomb. The calorimetric value of double base propellant lies in the range of 700–1000 cal g^{-1}.

7.2.7   Burn Rate Measurement

Propellant burn rates are measured using strand burners or acoustic emission techniques. A known length of sample is initiated from one end and the time taken by the ignition surface to reach a given probe location in the sample is measured to estimate the propellant burn rate. Acoustic emission based techniques have also been implemented for burn rate measurement of propellant strands.

7.2.8   Sensitivity Tests

The sensitivity of propellant for impact, friction, shock, jolt, vibration and other external stimuli are also important QC parameters. These parameters are mandatory for assessing hazards in manufacture, handling, storage and transportation of propellants. A BAM fall hammer apparatus imparts impact energy to the sample by means of a falling weight. The limiting impact energy is determined as the lowest energy at which a flash, flame or explosion is observed. The test is used to assess the sensitivity of the test material to drop-weight impact. The BAM fall hammer test is part of UN Test Series 3, which is used to assess the ignition sensitivity of suspected explosive materials (Figure 7.5).

The friction sensitivity test is used to measure the sensitivity of test materials to frictional stimuli. The test is a part of UN Test Series 3, which is used to assess the ignition sensitivity of suspected explosive

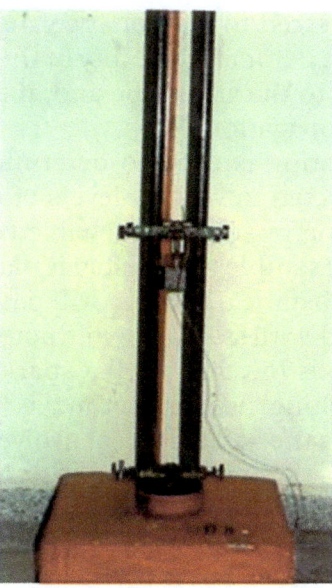

Figure 7.5 A fall hammer apparatus.

Figure 7.6 A friction sensitivity tester.

materials. A 10 mm³ sample is spread on a porcelain plate and the plate is then dragged under a weighted porcelain peg (Figure 7.6). The force on the peg is varied and the limiting friction load is determined as the lowest force for which a flash, flame, or explosion is observed.

Detonation shock testing is used to assess the detonation ability of a material. The sample material is loosely filled into a 0.5 m long steel tube having a 50 mm internal diameter and 60 mm outside diameter.

The tube is then subjected to a detonative shock from a high explosive donor charge. The detonation ability of the sample is determined based on the damage to the steel tube and, if necessary, by measurement of the rate of propagation.

Spark sensitivity testing is used to determine the response of an explosive, when subjected to various levels of electrostatic discharge energy. Electrostatic energy stored in a charged capacitor is discharged to the test sample. The sample to be tested is placed on a special holder that assures the electrostatic discharge will pass through the sample. A capacitor is charged with a known volt potential (usually 5000 V). The discharge needle is lowered until a spark is drawn through the sample. The approaching needle method is most commonly used because it best models the safety issues involved with ESD sensitivity. An infrared analyzer is normally used to determine sample initiation. A typical apparatus is shown in Figure 7.7.

7.2.9   Thermal Analysis

Differential thermal analyzer (DTA), thermo-gravimetric (TG) analyzer, differential scanning calorimeter (DSC) are used for sensitivity measurements of propellants towards thermal shock and ignition temperature of propellant compositions and ingredients. DTA (Figure 7.8) measures the behavior of a substance at a controlled rate of heating against a reference. TG operates in the temperature range from ambient to 1100 °C with programmed heating of the sample up to a

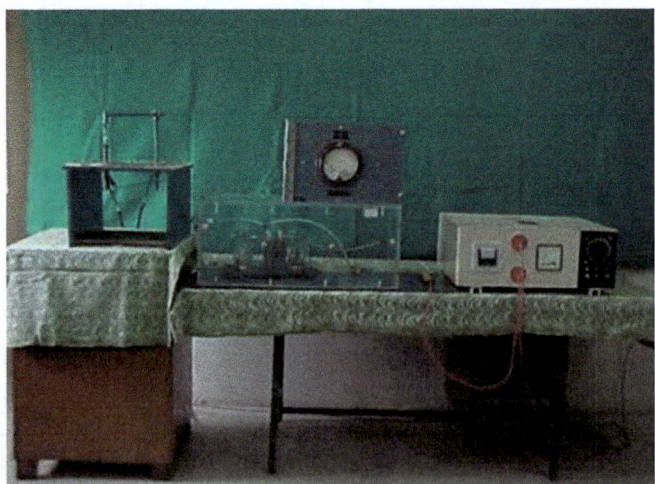

Figure 7.7   A spark sensitivity tester.

Figure 7.8 DTA apparatus.

scan rate of 100 °C min^{-1}. It is helpful in thermal decomposition studies and can give decomposition products directly, if coupled with a
Fourier transform infra-red (FTIR) spectrophotometer. DSC operates
with a programmed heating/cooling rate of 0.1–200 °C min^{-1} (in 0.1 °C
increments). The power compensated temperature null principle
based DSC measures the energy of transition directly. It also offers a
means to determine the phase/physical state transition temperature
including glass transition/melting/crystallization temperature, heat
capacity, purity and kinetics of the materials.

7.2.10 Testing Mechanical Properties

Testing the mechanical properties of a solid propellant is one of
the important mission requirements. Tensile strength, compressive strength, percentage elongation, modulus, Poisson's ratio, bulk
modulus, and visco-elastic properties are specified in mission specifications and compliance to these specifications is a mandatory
requirement for propellants. A universal testing machine (UTM) has
been used for almost all the process plants to ascertain mechanical
properties data. A sample in a double-dumbbell shape is conditioned
at a given temperature and is fixed in the jaws of the UTM (Figure
7.9). The sample is elongated at a given strain rate and stress *vs.* strain
diagram is obtained for the propellants, which gives tensile strength,
percentage elongation and modulus. Similarly, for compressive strength,
cylindrical samples are taken. The visco-elastic characterization of
the propellants needs a dynamic mechanical analyzer (DMA), where

Figure 7.9   UTM.

the frequency response is given as the input and temperature time superimposition is applied to ascertain the value of tan delta, loss modulus, and other parameters.

7.2.11   Radiography

For propellants, radiography has been the most suitable method for non-destructive testing (NDT). The principle of radiography is based on revealing optical density and thickness variations of X-rays, while passing through a sample and impinging on a screen (Figure 7.10). Real time radiography (RTR) is the latest addition to such a technique, where X-rays coming out of the sample are converted to extremely low intensity electrons and is displayed on a video monitor after acceleration and multiplication. Depending on diameter of the grains, the materials used for making propellant and internal contours, the intensity of X-ray machine is selected. 50–400 keV are generally found to be suitable for inspection of metal thicknesses up to 7.5 cm steel equivalent. For higher thicknesses, linear accelerators up to 20 MeV are used. Iridium-192 and cobalt-60 combined together can cover inspection ranges of 10–200 mm of steel equivalent. In RTR, CSA (NA) is used as a fluorescent layer for X-rays below 400 keV, while hard radiations are achieved with rare earth phosphors with some

Figure 7.10 Radiography set-up for propellants.

loss of sensitivity. With the advancement in science, high energy X-ray sources are being manufactured. Right from resonant transformers (1 MeV), Van de Graaff generators, Betatron and now high power Linear Accelerators (LINAC) have given us enough tools to decipher the interiors of propellant grains with higher accuracy and clarity. Depending of type of source and sensitivity requirements, radiographic density, effective diameter of focal spot, geometrical sharpness, attenuation of source and contrast of developed film changes, it is an art rather than a science to correctly interpret the results from these radiographic studies.

7.3 Instrumental Techniques

7.3.1 Chromatographic Techniques

Thin layer chromatography (TLC) is the oldest chromatographic technique but is simple, rapid, inexpensive and relatively sensitive. The combination of the distance traveled by a compound and the color reactions used for the detection of compounds has been used for the identification of several important ingredient of propellants. TLC has been used for identification of NC by acetone–methanol (3 : 2) as eluent, NG by toluene–cyclohexane (7 : 3) eluent *etc.*

High performance liquid chromatography (HPLC) is used for the detection of trace metal elements and also metallic salts. This method allows the analysis of stabilizers like diphenyl amine (DPA), 2NDPA, methyl centralite, ethyl centralite and plasticizers like diethyl phthalate (DEP), DNT, diethylene glycol dinitrate (DEGN), NG, metriol trinitrate *etc.* In many cases the analysis of all the extractable ingredients in a propellant can be performed simultaneously.

Gas chromatography (GC) is very useful in the identification and estimation of toxic gases that are normally present in the combustion products of gun and rocket propellant compositions. GC is mainly used for thermally labile and volatile compounds. The vapor of the sample is injected into a column containing the stationary phase. The sample is pushed through the column by a carrier gas, which constitutes the mobile phase. When a mixture of different components having different interaction with the stationary phase is introduced, the components move through a column at different rates and emerge at different retention times. This way separation is completed. Elements are identified by detectors such as a thermal conductivity detector, a nitrogen phosphorus detector or a thermal energy detector.

7.3.2 Spectrophotometry

Atomic absorption spectrophotometry (AAS) is used for the detection of trace metal elements and also metallic salts. It is an analytical method for the determination of elements based upon the absorption of radiation by free atoms. The combustion flames provide a simple means of converting inorganic substances in solution into free atoms when introduced in the form of an aerosol. A large number of these free atoms remain in an unexcited state or in the ground state. The vaporized atoms are then exposed to radiation from a light source. It emits only those frequencies of light that are present in the emission spectrum of the atoms. Likewise, the vaporized atoms in the flame will only absorb these frequencies when it contains this very element. Once the radiation has passed through the sample, a monochromator, consisting of a diffraction grating and a slit isolates the desired radiation frequency and transmits it to a detector. The instrument is used for the estimation of metal oxides (burn rate modifiers) present in the propellant. A UV spectrophotometer is used to find the maximum wavelength before the compounds are taken for analysis by HPLC apart from identification of organic and some inorganic compounds.

7.3.3 Memotitrator

This is an electrometric titrator, used for the quick determination of moisture content and titration of mixed acids. If the end point in a titration is not clear, due to either highly colored solutions or lack of proper indicator, this technique can be adopted. It is used for the estimation of ammonium perchlorate in the propellant samples and nitric acid, ammonium nitrate and acetic acid in the spent acids. It has also been used for oxidation–reduction as well as precipitation titrations.

7.4 Calibration of Equipment

QC requires calibration for all the equipment used. Calibration is the process of determining deviations in indicated values of the measurements by the equipment and values measured by a standard. Standards are kept and reproducibility of their readings are checked globally by key comparison. Calibration only indicates deviations of the given equipment in measuring a limited range of parameters. For example, a curing oven has to maintain a temperature and thermocouples are installed with the oven to indicate this. For the calibration of the thermocouples, a standard hot water bath is required, whose temperature is constant. The temperature of this bath is measured with the help of a thermocouple, under calibration and a standard thermocouple and the difference of readings are indicated in the calibration report of thermocouple along with the range of temperature in which calibration is completed. Calibration has limited life because of repeated use and misuse of the equipment, variations in environmental and operating conditions, wear and tear of anvil *etc.* All the equipment used in measurements must be calibrated and should have repeatability and reproducibility of values.

7.5 Ballistic Evaluation

Since the flight testing of propellants for lot proof is expensive, an evaluation of propellant performance is ascertained by static evaluation. Static evaluation of propellant grains is conducted in various stages. With each lot of large diameter propellant grains, a carton grain is also processed, which undergoes a similar processing cycle as the actual grain being tested is undergoing. The propellant samples for mechanical properties evaluation, burn rate determination by strand burner, density *etc.* are generally obtained from these grains. From similar grains ballistic evaluation grains are also obtained. Ballistic evaluation motors are very popular for such assessments, especially for evaluation of ballistic properties. They give an insight into the burn rate of the propellant at the operating temperature and pressure ranges and help in calculating characteristic velocity and

Isp. A popular preliminary ballistic evaluation motor (BEM) contains a tubular propellant grain with no inhibition. It burns from all sides and gives a slightly regressive pressure time profile. The propellant quantity involved is around 2.5 kg and has dimensions of outside diameter of 115 mm, an inside diameter of 60 mm and length 200 mm (Figure 7.11).

A slight regression in the profile results in covering a large pressure range and a burn rate law over a pressure range can be obtained. These are used for research activities mainly and a large number of compositions can be evaluated in different pressure ranges (obtained by changing throat area) by incurring minimum cost. In France, a propellant grain code-named as "CAMPANULE" is used for such activities. This has a 10-point star port with 90 mm outside diameter and 300 mm length. This weighs around 2.5 kg. In other countries, cylindrical grains are preferred. Another class of propellant configuration, which has been extensively used in India for BEM, is called 40 kg BEM grain or "AGNI" motor. These grains are generally configured as star shaped with a recess or port at the centre. In France, it is customary to use star port grains with 10 petals as a control solid propellant grain for the ballistic properties evaluation of propellant batches. It weighs around 45 kg with dimensions of 203 mm diameter and 1 m length. It shows good neutrality, defined as percentage variation in the surface area of propellant during burning. In the US, BATES (Ballistic Test Evaluation System) type of grains are very popular. It is reported to give constant burn area *vs.* web burnt profile with no combustion tail-off. For specific situations like for the characterization of energetic binder compositions, a very good tail-off is desired so that any unburnt residual propellant can burn with other initial surfaces, but for BEM motors the least tail-off is desirable.

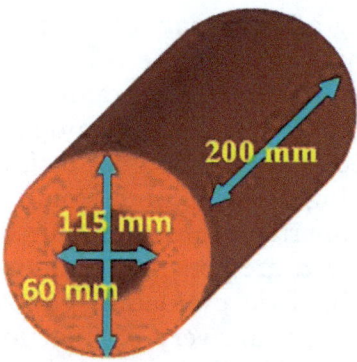

Figure 7.11   A BEM propellant grain.

Ballistic evaluation needs a static test bed of adequate capacity (5–100 T). It needs a thrust block/bed against which rocket motors on a trolley is butted (Figure 7.12). The instrumentation for the measurement of ballistic and performance parameters like pressure, thrust, strain, burn time, skin temperature, displacement, vibration, sound level, and high speed videography along with all the safety features are available. The head end of the motor under a preload is butted against the thrust block and the rocket motor is fired remotely. The measurements, done at a remotely situated control room give the complete performance parameters in terms of pressure–time profiles, thrust–time profiles (Figure 7.13), characteristic velocity, Isp, burn rates at different pressures and strain variations *etc.* A test bed with advanced instrumentation and conditioning facilities has always been part of QC as well as an essential feature of any propellant production plant.

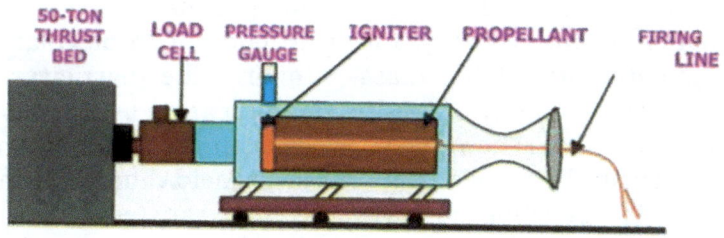

Figure 7.12   A static test bed set-up.

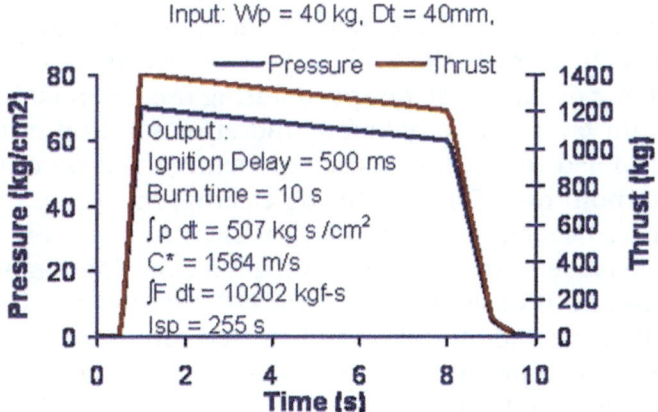

Figure 7.13   A static evaluation curve.

7.6   QC for the Manufacture of Composite Propellants

Presently, composite propellants have been used in most missiles and space vehicles. They are obtained by mixing an oxidizer, fuel (polymer and metallic), plasticizers, burn rate modifiers, bonding agents, process aids and curing agents *etc.* in the form of a slurry. The slurry is cast under vacuum in a mould around a suitably shaped mandrel and cured at an elevated temperature. The solidified propellant grains are demoulded, cored, machined, inhibited and radiographed and stored/evaluated. Demoulding is dispensed with in case bonded rocket motor manufacture. For QC during the manufacture of composite propellants, both the ingredients and process parameters are controlled.

All the solid raw materials (oxidizer, AP; fuel, Al) are checked for purity, moisture content, particle size and volatile matter. The liquid ingredients are generally monitored more strictly. The workhorse binder HTPB is tested for hydroxyl value (35–40), acid number (<2), functionality (~2.3), viscosity, moisture content, molecular weight, specific gravity and volatile mass. Similarly, the plasticizer DOA is tested for specific gravity, viscosity, saponification (~306), acid value and refractive index. The process aid lecithin is tested for moisture content, viscosity, benzene insoluble matter, acid value, acetone insoluble matter. The bonding agents (Mat-*O*-Bond) are tested for hydroxyl value, acid number, volatile matter and moisture content. The curing agents (TDI, IPDIHMDI) are tested for assay, hydrolysable chlorine, specific gravity, refractive index and isomer contents.

Each of the ingredients is weighed on calibrated balances, which has sufficient accuracy for the measurement range. Mixing is a major process, which controls the end properties of composite propellant grains. Mixing duration, blade speed during mixing, mixing temperature, vacuum level, viscosity before and after addition of the curing agent and moisture content in the mix before addition of the curing agents are monitored for QC. During casting, the vacuum level, mix temperature, viscosity build-up and casting rate are monitored. The curing cycle is the next operation, which is a combination of time and temperature, both of which are major QC parameters. After curing, the propellant is ready for use and is tested for density, calorimetric value, mechanical properties and chemical composition along with a visual inspection. With the main propellant grains of a lot, carton grains and BEM grains are always cast. Ballistic evaluation is carried out to ascertain the ballistic competence of the propellant produced. X-Ray analysis is mandatory for propellant grains before evaluation. For cartridge loaded propellant grains, 450 keV X-ray machine is sufficient but for large caliber grains and mainly for case bonded applications a minimum 6 MeV X-ray linear accelerator machine is needed for radiography. Case bonding brought more load on X-ray machines, as in addition to grain radiography, exposures of interface becomes more important and number of shots for clearance of motor has increased tremendously for such applications. A typical QC flow diagram is shown in Table 7.1.

Once cleared, the composite propellant grains/motors are ready for inhibition, followed by re-radiography and static evaluation/flight evaluation/storage for future use.

Table 7.1 QC checks and their importance.

Processing events	QC parameters	Importance
Processing of raw materials	Particle size, moisture content, purity, volatile matter *etc.*	Burning rate, porosity
Mixing	Sequence of addition, temperature, viscosity, vacuum	Homogeneity, pourability
Casting	Mould assembly, viscosity, vacuum, temperature	Castability, core extraction, voids
Curing	Temperature	Thermal stress, mechanical properties *etc.*
NDT	Cracks/voids	Performance variation
Trimming/ inhibition	Dimensions, bond strength	Irregular burning
Grain/motor inspection	Mechanical properties, burn rate, sensitivity	Prediction of ballistic performance, safety

8

Process Safety

8.1 Introduction

The rocket and propellant industries are faced not only with normal occupational hazards, but also with the necessity of working with materials that are combustible and explosive in nature, as well as being toxic and corrosive. The presence of some materials presents a hazard not only to civil work and equipment but also to the worker. Safety therefore is a must for survival, growth and prosperity. Besides many technical factors, there are certain common psychological factors that must be treated before any scientific and technological upgrade. Social prosperity, security and survival without any ill effects must be achieved.

Safety conjures up thoughts of accidents, injuries, compensation and investigation. The causes of unsafe acts are attributed to carelessness, ignorance, sabotage and negligence and at times passed off as an act of God. Despite statutory measures for protection and heavy penalties for violation, the practice that is being followed is to correct the specific instance that occurred.

Explosive safety is one of the most difficult and demanding tasks due to the requirement of striking a balance between operational and explosives safety considerations. The advances in explosives and weapons technology must be viewed in conjunction with safety engineering and research and development testing. There is a need to understand the ability and necessity to identify hazards and design

Solid Rocket Propellants: Science and Technology Challenges
By Haridwar Singh and Himanshu Shekhar
© Haridwar Singh and Himanshu Shekhar 2017
Published by the Royal Society of Chemistry, www.rsc.org

or engineer them out or at least control them. Explosive safety has been evaluated by accidents that have occurred in the past. All these experiences have been condensed into rules and laws, describing what is proper and what is not. The modern approach in nuclear and chemical safety regulations is to set the general legal goals for safety and leave the means of achieving them to be determined by the experts. Military research not only aims at destruction but also at protection of soldiers, equipment and facilities against the effects of weapons.

In the process of the production of rocket propellants and explosives, a number of safety requirements have to be met to eliminate the ignition source. It must be understood that it is a function of explosives to explode with different types of stimuli. However, explosives should detonate, when they are expected to do so and should remain safe/stable during processing, storage, handling and transportation. The explosive is a property of any dynamically excited two-phase system. Explosions have taken place even with black powder over the centuries. Alfred Nobel experienced all possible failures with nitroglycerin and nitro-cellulose. During the last 100 years there have been several major accidents in many countries due to fire, explosion and accidental dropping of munitions. In NATO countries more than 6000 people have lost their life and 16 000 people fatally injured. About 8000 acres of property were lost in these accidents. Table 8.1 shows an analysis of the causes of such accidents.

It can be seen from Table 8.1 that the majority of accidents took place with pyrotechnics and other pyro devices, followed by explosives. Comparatively lower numbers of accidents have taken place with propellants. The casualties and injuries were higher with pyrotechnics. The most dominant factor for accidents is friction, in all cases.

Safety is represented by a green triangle. We never fix the guiding factors on the three arms of this triangle; environment, enforcement

Table 8.1 Analysis of causes of accidents.

Stimulus	Pyrotechnic (%)	Propellant (%)	Explosive (%)
Impact	6	10	13
Friction	66	47	65
Spark	5	2	6
Heat	24	40	15
% incidents leading to injury	31	8	19
% incidents leading to death	3	0	1

Figure 8.1   The safety triangle.

and operation (Figure 8.1). Any weakness in any part of the three arms of the safety triangle is going to break, thereby leading to calculated or unimaginable consequences. In addition to technical studies and technological upgrades, certain values are set by all three arms of the safety triangle like operation and workers, employer and management enforcement/regulatory authorities. It must be clearly understood that safety is everyone's responsibility, not only in the explosives industry but also in every walk of life.

Sometimes there are fatalistic views of life *e.g.* parental messages like what is going to happen, will happen? What is slotted can't be blotted! Events are always predetermined! We are pawns in the hands of destiny. Anyone and everyone has to accept that accidents are totally avoidable and absolute safety is well within our reach, provided a firm mental determination exists. Some of the important factors for accidents include:

- Lethargy, haste and detouring—human factors of lethargy and haste along with mental detouring dominate the work psyche, which leads to accidents.
- A sense of complacency—the over dependence on auto-controls that will take care of everything brings complacency among operators, which renders them non-vigilant, non-alert and consequently leads to accidents.
- A sense of excellence with half knowledge—a contributory factor behind many unsafe acts and practices.
- Environment—a clean environment is mandatory.

Technological advancements have contributed to a low probability of accidents in explosive activities through the production of generally safe explosives, continuous QC during manufacturing and storage, rigorous safety rules for the personnel involved, early warning systems and protection devices and rigorous security protection. The new safety concept provides a detailed analysis of the frequency of accidents and quantification of the risk. This includes:

- A realistic analysis of the explosive effect.
- A realistic prediction of the people present in a hazardous zone.
- A realistic probability of accidental explosions.
- Explicit criteria for acceptable risks.
- A cost-benefit analysis.

Studies have shown that human error is the cause in some of the 90% of the incidents and that 70% of the incidents could have been prevented by management action. This points to the crucial significance of a systematic approach to the management of health and safety and the need to be aware of human factors as a distinct element in that framework. A problem, which is unique to the HEM industry, is that although the hazards may be appreciated, the low probability of an event may endanger complacency. An important goal is to secure better recognition and understanding of effective safety management.

Training (fire drill and safety awareness programs) is an essential ingredient of any successful safety policy. The lack of training is a major contributory source of human error. National safety week is observed once a year in the first week of March (4th to 10th March) in almost in all organizations to rededicate the workforce towards safe working practices. However, it should be practised every minute at work.

8.2 Classification of Hazard

The United Nations (UN) classification of hazardous materials has nine hazard divisions: explosives, gases, inflammable liquids, inflammable solids, poisonous substances, radioactive substances, corrosive substances and miscellaneous materials. Propellant comes under UN classification 1. This classification is further subdivided into six headings:

1.1. This comprises of materials prone to a mass explosion hazard. Major hazards by substances of this class include blast, high velocity projection and flame. The structural damage caused is directly proportional to the quantity and type of material involved. Initiatory, high explosives, mines and propellant mixing are classified here.

1.2. Substances in this category have a projection hazard and a minor explosion hazard but not a mass explosion hazard. Burning followed by progress towards explosion in small quantities is observed. A considerable number of fragments, firebrands and unexploded items may be projected. These projected pieces can explode or detonate on impact, but the blast effect is limited to the nearby vicinity. Assembled munitions, grenades, and rockets are classified here.

1.3. Finished propellants are classified in this hazard division, which comprises items having a mass fire hazard. They may have a minor blast hazard or a minor projection hazard but not a mass explosion hazard. They may burn with great violence and intense heat resulting in dangerous fragment formation or a thermal radiation hazard.

1.4. This division includes items that pose, primarily, a moderate fire hazard. These substances do not contribute to fire or the formation of fragments of any appreciable sizes. The hazard is largely confined to the packages. Small arms ammunition and caps are included here.

1.5. This group was created for very insensitive substances that have a mass explosion hazard. There likelihood of initiation or transition from burning to detonation under normal conditions of transport is negligible. No military explosives are classified in this group.

1.6. This group contains substances that have the chances of an explosion limited to a single article. No mass explosion hazard, negligible probability of accidental initiation or propagation is expected from these materials. No military explosives are classified in this group.

Each of these heads has a typical symbol as depicted in the Figure 8.2. Four figures containing numbers in different red-colored geometries inside a square indicates divisions 1.1 to 1.4.

Fire is one of the major hazards during processing, storage, transportation and end use of the propellant. The fire needs an oxidizer, a fuel and a heat source for initiation and propagation. In addition

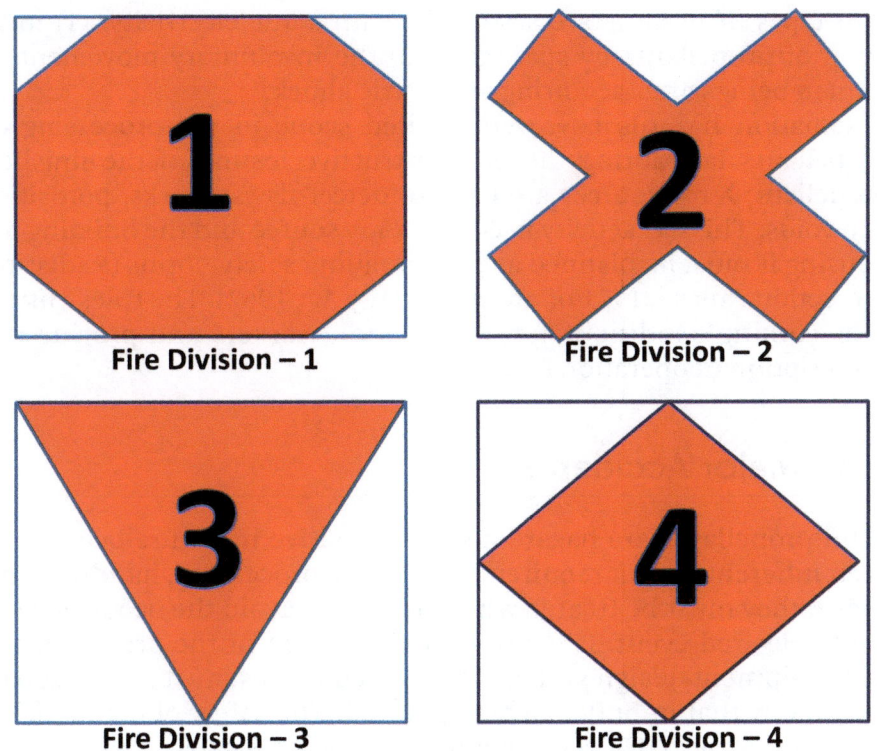

Fire Division – 1

Fire Division – 2

Fire Division – 3

Fire Division – 4

Figure 8.2   Fire division symbols.

to this, it also needs the continuous formation of reactive groups for sustenance. Fire is grouped into four classes. A class 'A' fire involves solid organic materials like wood, paper, textiles, rubber *etc.* A class 'B' fire includes liquids and liquefiable solids like alcohol, petroleum, oil, paints *etc.* A class 'C' fire exists in the presence of gases and liquefied gases in the form of a liquid spillage or a gas leak as occurs for hydrogen, propane and acetylene. A class 'D' fire involves metals and their alloys like potassium, sodium, aluminium zirconium *etc.*

Static electricity is an invisible silent killer. When two dissimilar non-conducting bodies rub against each other, an electric charge at the interface of the bodies gets disturbed, resulting in one body acquiring a higher positive charge at the cost of a raised negative charge on another body. When the accumulated 'static charge' exceeds the breakdown potential of the medium, it results in a high voltage, low current discharge in the form of a 'spark'. The ignition energy of a sensitive propellant composition may be as low as 0.001 J and the accumulated charge can have sufficient energy to ignite the propellant.

Two types of hazards are associated with static electricity: (1) initiation of propellants by spark and (2) the involuntary movement of human beings after acquiring an electric shock.

Radiation hazards have very limited scope in the processing of propellants but during the non-destructive testing of the finished propellant, X-ray sources are used to detect flaws, cracks, porosities and voids. This QC activity needs an X-ray source and the building for housing it must be planned as per prevailing safety norms (Radiation Protection Rules – 1971 of Atomic Energy Act 1962). The rules ensure safe working conditions for all radiation workers and provide the prescription of operational dose limits.

8.3   Major Accidents

Solid propellants are hazardous and sensitive in nature because of their inherent output requirement. During processing, handling and usage, they must be treated with due care to avoid the occurrence of any undesired event. It is not only the design of the process plant but equipment design and processing sequences must be devised in such a way that in-built safety is ensured. The safety of personnel is of prime concern. However, plant safety, process safety and material safety data are by no means any less important. The hazard and risk need to be tamed and it is ideal if 'zero accident level' is achieved by keeping the frequency and intensity of accidents to the barest minimum or zero.

Although accidents involving materials lying in UN hazard division 1 have been reported, most of the accidents took place whilst processing, storing and using pyrotechnic devices. High explosive ingredients filled in warheads come next followed by propellants. Leaving aside, accidents involving ammunitions, bombs and explosives, a brief account of accidents involving propellants alone is summarized below.

On 12th March 1907, the accidental ignition of gun propellants in the ammunition magazine in France resulted in a fire explosion killing 117 and injuring 33 people. In a similar incident on 25th September 1911, 233 people died and 160 were injured. During both World Wars and the following 40 years, several ammunition depots, ammunition carrying trains *etc.* were destroyed by accidental initiation of explosives in the USA, USSR, UK, Japan, France, Germany, Vietnam, Cuba *etc.* On 11th January 1985, a Pershing II motor exploded in Germany, whilst being lifted from its shipping container, killing three people and injuring nine. It occurred due to an electrostatic discharge coupled with sensitivity enhancement of the propellant in the cold, dry weather. On 27th December 1987, a propellant ignited whilst decoring from an MX motor (solid propellant rocket motor) killing five people and destroying the facilities at Utah (USA). The cause assigned was electrostatic discharge and friction. Ammonium perchlorate, a key ingredient of modern solid propellants, burned and detonated killing two people and injuring more than 350 local residents. The event took place in Nevada, USA and the quantity of ingredient involved was approximately 8 million pounds. The event in Utah was repeated in India on April 2002 killing six people and another eight people in November 2003.

In addition to processing and storage, the usage of solid rocket propellants in various systems also require proper care. The fates of several static evaluation motors, which result in bursting, all over the globe go unnoticed. The reasons for this are generally improper propellant loading, inner grid failure, inadequate thermal insulation, nozzle blockage, poor mechanical properties of propellant and motor casing joint failure *etc.* But several failures in flight conditions have been reported in the literature. On 2nd August 1993, a Titan IV SRM of the US resulted in a burst, after 101.2 s against an expected burn time of 127 s. The failure was attributed to restrictor repair for which a cut 0.25" deep and extending 34" in the radial direction in the propellant was drilled. Under operation, the cut was supposed to close but it opened resulting in the early exposure of the casing to the hot combustion gases. On 18th April 1986 a Titan 34D-9 SRM resulted in a burn through of the motor case due to unbonded insulation. On 28th January 1986, an STS Challenger (51-L) resulted in failure due to the early exposure of the metallic casing to the combustion flames. A recess for the flow of gases was created due to a loss in flexibility in the O-rings under the cold operating conditions. On 12th September 1985, Ariane 3, launched from Kourou, French Guiana, stopped operating 277 s after launch against a normal burn time of 720 s. This failure was attributed to the malfunctioning of the cartridge-type igniter in the combustion chamber.

Table 8.2 compiles a brief description of some of the fatal accidents (causality 'C', serious injury 'SI', injury 'I') that have taken place in India. This table does not include accidents involving loss of property, minor injuries or data from the civil explosive sector.

Table 8.2   Fatal accidents in India.

Year	C	SI	I	Cause	Remedial measures
		Fatalities			
1992	2	2	3	Arming of warhead during assembly	Ensure the safe practice of disarming the warhead during assembly and the use of a safe detonator (non-sensitive)
1992	4	—	1	Increase in temperature during processing of ball powder (NC + 8% NG)	Additional temperature, pressure gauge and pressure release rupture devices to be included
1992	10	—	—	Contamination of coated Mg powder with water, resulting in ignition and dust explosion	Moisture content to be monitored rigorously. Effective earthing of all equipment to be implemented. Al or SS containers to be used for Mg
1994	9	3	—	Accidental fall of heavy object on loose composition of primers	Immediate removal of rejected primers and storage under water. Ensure explosive store limits for buildings and operations
1995	3	—	1	Presence of dried NC dust in vibrator assembly	Make operations remote controlled. Improve deluge system. Provide additional sensors. Volatile matter not to be less than 30% in NC cakes
2000	1	—	—	Auto decomposition of propellant in old and unused ammunition	Disposal action for old and unused ammunition to be undertaken periodically
2001	2	—	—	Inadvertent ignition of assembled rocket motors	SOP not followed
2002	6	—	3	Friction during extraction of propellant. RH very low	RH control very important during peak summer. Multiple escape routes in extraction bay. Only one activity to be carried out in the building during extraction

Major cause of various reported accidents are as follows:

- Carelessness.
- Over confidence.
- Non-control of moisture/volatile matter.
- Non-adherence to stipulated standard operating procedures (SOPs).
- Multiple process activities at a time.

The real cause of the explosion can be impact, friction, self-heating, influence of shock wave, radiation, electrical discharge *etc.* The first stage in a chain of events leading to energy evolution, which results in destruction and demolition of environment is *always the combustion of high explosives.* Correspondingly, every accidental explosion in some sense might be considered as a consequence of a fire. Primary explosives detonate at ignition. Secondary explosives need a blasting cap filled with primary explosives to detonate reliably and effectively.

In view of above criteria for heat sensitiveness a classification has evolved. Very sensitive explosives are categorized under 'A' with an ignition temperature less than 200 °C. Category 'B' contains high-energy materials with ignition temperatures between 201 and 300 °C. Composite propellants with an ignition temperature of 250 °C fall within this category. Category 'C' is for comparatively insensitive materials having ignition temperatures of greater than 300 °C. The thermally stable explosive TATB falls within this category.

8.4 Safety During Propellant Processing

All operations during propellant processing must be supervised by an independent safety representative, who has to ensure the safety during processing as per the shop instruction of GSD. The safety representative should prepare a checklist for ensuring this. The various features of a typical checklist are shown in Figure 8.3. This checklist must be filled in before starting any new operation in the building. In addition to this, there are structured visits of senior officers and safety reviews enhances the awareness towards safe working practices and ensures accident free operations on the site.

Propellant mixing comes under class 1.1 and carries a major explosion hazard. Mixing must be conducted in a traversed building with remotely monitored feeding and operation controls. The propellants

Check Points

1. Building No. and name
2. Traverse
3. Roof
4. Manpower limit
5. Explosive limit
6. Floor
7. Activity
8. Present safety monitoring
9. escape door
10. waste disposal
11. log-book maintenance
12. Doors
13. Electrical Safety
    a. Lightning protection
    b. Earthing
14. Drainage
15. Standard operating procedure (SOP)
16. Tooling
17. Water tap
18. Static tank
19. Fire hydrants
20. House keeping

Figure 8.3 The common checklist points.

are mixed in either horizontal sigma blade mixers or in vertical planetary mixers. Sigma blade mixers are slowly being phased out due to exposure of bearings/glands to the propellant mix, interrupted mixing during ingredient loading, and limitations on the restricted batch size. Whatever may be the mixer type, the mixing of a highly filled polymer matrix of propellant needs physical movement of the entire mass during mixing by the blades of the mixer. A complete sweep inside the vessel requires a very low gap between the mixer blades and between the blade and vessel wall. Any accidental falls of foreign material can result in a fire hazard. In general, the ingredients are loaded in the mixing vessel through a sieve, to avoid any such event. In addition to this, the mixer blades are designed to move in only one direction and reverse sweep is generally avoided. Water deluge and bowl drop provision, in case of fire, are essential features of the mixing area. The exposure of the fine dust of the solid ingredients and toxic liquid ingredients can be avoided by using a close ingredient feeding system.

Drilling and propellant cutting are other critical operations and they require proper remote-operated machines with several safety precautions. Since the interaction of scientist and machine requires proper synergy, simple limit switch provisions and emergency shut-down switches must be provided. In addition to this, the use of non-sparking tools must be encouraged.

Safety against static electricity is possible through an efficient earthing of the working personnel and machines used for processing the propellants. The prime cause of static electricity is voltage difference, which is reduced to zero by bonding both materials by conducting wire followed by earthing of this wire. The sse of non-conducting materials in the work place must be avoided as far as possible. Synthetic clothes, non-conducting flooring and non-conducting shoes must be avoided while working with propellants. Non-conducting flooring is preferred for buildings where materials with an ignition energy less than 45 mJ are being processed. Anti-static flooring is relatively inferior in preventing electrostatic hazards because its resistance changes rapidly with environmental conditions like humidity, temperature *etc.* Dry conditions favor the accumulation of static charge. A person (body capacitance ~100 pico Farad) wearing synthetic clothes and non-conducting shoes can build up a voltage of the order of 25–40 kV during normal activities like walking, moving their hands *etc.* These motions correspond to an electrical energy of 165 mJ accumulated on the body.

In modern rocket motors, the hazard potential has increased. Many of the hazard tests used 20 years ago to discriminate between pro-pellants and explosives are no longer adequate because of the use of nitramines (RDX/HMX) with NC/NG systems. Performance will always continue to be major consideration. Hence, the need for performance/hazard trade-off exists. Past methods for ranking solid propellants are inadequate. The classification of propellants was separate from those for high explosives and did not recognize the fact that some of the solid propellants and high explosives have almost the same combination of ingredients (AP/Al/RDX/HMX).

Advanced solid propellant safety is important (and significant) because of the potential damage with highly loaded energetic materials and the impossible task of precluding their inadvertent initiation. Past incidents are relevant to present and future incidents that could occur with modern solid propellants. There is an increased concern for survivability of weapon platforms in a combat environment and the desire for 'insensitive munitions (IMs)'. It is easy to declare the desire for IMs but it is more difficult to design them.

8.5   Common Dos and Don'ts

Some common dos and don'ts in the propellant processing area are given below.

Do:

- Keep the floor area, shelf and work area clean and tidy. The spillage of ingredients, tools, devices, jigs and fixtures should be avoided.
- Keep fire buckets full with sand and water. Always keep a suitable fire extinguisher ready for use at a convenient nearby place.
- Use personal protective equipment like nail-less boots, aprons, ear and nose masks, safety goggles *etc.*
- Ensure that all machines in the processing area are properly earthed.
- Keep doors and windows open whilst working (wherever possible).
- Discharge yourself before entering the building (always).
- Check the lightning arrestor and electrostatic discharge system periodically.
- Shut down the power supply after completing a job and ensure that the buildings are closed afterwards.
- Ensure that the water deluge/sprinkler system is on during propellant operation.
- Ensure the capacity of slings, D-shackles, cranes and handling devices during operation.
- Ensure that the feeding hopper at the feed ports is fitted with a sieve during handling powdered raw materials.
- Remove the silica gel bags from containers after opening drums before charging powder for the mixing and grinding.
- Use cotton bags and overhead cranes when loading material.
- Unload materials in properly labeled drums.
- Ensure that the oxidizer and burn rate modifiers are dried in separate ovens.
- Ensure that any spillage of the casting liquid is wiped with diethyl phthalate (DEP) wetted cotton waste and then cleaned with solvent.
- Always run the mixer empty to ensure its proper functioning before starting the mixing program.
- Circulate the water in the vacuum pump before it is switched on.
- Check the mixer bowls for cleanliness before loading.
- Check the water level in the water-jacketed ovens whilst curing.

- Ensure that the temperature shown on the control panel and the actual temperature in oven is the same before starting curing.
- Switch off the oven before keeping the charge inside the oven.
- Keep moulds and rocket motors inside the oven securely and ensure that there is sufficient space for handling.
- Switch off the oven during other operations in the building.
- Use asbestos hand gloves for the handling of hot moulds.
- Switch off ovens after curing and remove the moulds and motors only after ambient temperature is attained.
- Earth the mould and core whilst decoring.
- Check the alignment of the ram of the hydraulic power pack during decoring.
- Keep the speed, feed and depth of cut as per regulations during machining.
- Ensure a continuous water flow on the work-piece and cutting tool for wet machining.
- Take only one charge at a time for cutting, trimming and other machining operations.
- Ensure that a water supply is available, while machining the propellant.
- Use clean cotton waste free from grit, propellant waste and foreign material.
- Collect all propellant dusts in a covered bucket.
- Flood the propellant charge with water if a spark is observed during machining.
- Ensure that the relative humidity of the process building is in the range of $55 \pm 10\%$. Lower Rh makes the propellant charges sensitive to friction and impact.
- Take a blank run of the machinery before actual use.
- Dispose experimental batches separately.
- Conduct the waste propellant burning using a long wire, battery and igniter.

Don't:

- Exceed manpower and explosive limits during various stages of processing propellant.
- Subject propellant charges to jerks, friction or impact during handling, processing, machining, transportation and storages.
- Smoke in the propellant area.
- Sieve the oxidizer (AP) without earthing the tray.

- Keep combustible materials, alkalis, solvents, and acids in the process room.
- Put on the mixer whilst cleaning. Remember to turn the mixer off during QC inspection checks.
- Run the mixer blades in both directions.
- Open the oven whilst it is on.
- Pack the oven with too many moulds.
- Keep raw materials for drying in the same oven that is being used for propellant slurry curing.
- Burn tubular pieces during waste disposal. Cut them longitudinally. The size should not exceed 300 mm.
- Carry waste propellant in loose conditions. Always use covered drums.
- Drag bags/drums on ground. Avoid friction.
- Burn waste propellant in the absence of a fire tender.
- Burn waste propellant in a heap. Spread evenly to form a bed.

9

Ignition Systems

9.1 Introduction

The ignition of a rocket motor is an important and vital event in the combustion of solid propellants. It is a transient phenomena leading to the steady-state combustion of the propellant. An igniter should provide the optimum energy to raise the surface temperature of the propellant grain from ambient temperature to its ignition temperature. It should pressurize the initial free volume of the motor to a pressure level well above minimum pressure required for the steady-state combustion of the propellant without exceeding the maximum expected operating pressure (MEOP) limits of the motor. The rate of pressurization should be smooth to avoid any pressure peaks and combustion instability. Any ignition delay should be within specific limits. Moreover, the igniter should be easy to assemble and economical to produce.

In a rocket combustion chamber, the ignition of liquid propellants can be effected by either hypergolic self-ignition or by external ignition. The use of hypergols is safe and simple but attaining spontaneous ignition in a very short time is a challenge. An adjustment of the ignition delay must take into account various factors like the chemical composition of the fuel and oxidizer, the ratio of reaction components, surface tension, intermixing, injection velocity and the temperature of the reaction components, reaction enthalpy, reaction-velocity and the gas pressure in the chamber. Contrary to this, a

Solid Rocket Propellants: Science and Technology Challenges
By Haridwar Singh and Himanshu Shekhar
© Haridwar Singh and Himanshu Shekhar 2017
Published by the Royal Society of Chemistry, www.rsc.org

non-self-igniting propellant needs external energy sources. This is achieved by pyrotechnic squibs mounted at the head end or by a high energy spark plug as in internal combustion engines or by a self-igniting chemical starting fuel (xylidine injected just before admission of gasoline) or by catalytic surfaces in the chamber (potassium permanganate saturated surface for hydrogen peroxide decomposition) or by auxiliary oxygen supply or by electrically heated glow plug. A short ignition delay and good combustion efficiency have been the driving force behind various research in this area.

Ignition involves the transfer of heat or thermal energy from some suitable source to the propellant until steady-state combustion is achieved. The time taken for this event to happen is called 'ignition delay'. In solid propulsion, energy is transferred from a pyrotechnic igniter, such as gunpowder or any other composition having high heat flux to the propellant surface.

In the family of solid propellants, while double base propellants (DBP) need comparatively lower energy for ignition, composite propellants require higher energy in terms of heat flux. DBP gets ignited with a small quantity of black powder (KNO_3 + sulfur + charcoal) of different sizes (G-12 to G-40) by means of electric match. The device, which contains the ignition material and match inside a container (aluminium), is called the igniter. Generally, energy is transferred to the propellant surface by all three mechanisms of energy transfer namely, conduction, convection and radiation. The impingement of hot solid particles from the burning igniter is effective in igniting the propellant surface. KNO_3–Mg along with binders are used for composite propellant ignition. These 'gasless igniters' have a high reaction temperature and produce hot particles, which transfer thermal energy by both direct impact and radiation.

The ignition of solid propellant rockets consists of a series of complex chemical processes. The production of hot ignition gases by igniter, transfer of energy to propellant grain exposed surfaces, decomposition of propellant to yield gaseous intermediates, flame initiation and propagation and finally exothermic reaction to produce self-sustained propellant combustion. The igniter design is critical due to the fact that ignition related processes have not been clearly understood and any mismatch in the systems may lead to ignition delay, pressure peaks, propellant chuffing, kinks or in some cases extinction of charges also. For a theoretical understanding of behavior of the igniters, two different theories have been proposed. As per solid phase or thermal theory, a solid phase reaction determines the site of first appearance of the ignition flame, while the gas phase theory

Table 9.1 Preferred igniter composition based on propellant type.

Type of propellant	Ignition temperature (°C)	Minimum stable combustion pressure (kg cm^{-2})	Type of igniter preferred
Double base (DBP)	165	35	Gun powder
Composite (CP)	240	15	Metal/oxidizer
CMDB	175	20	Metal/oxidizer

concentrates mainly on chemical kinetics appearing in the combustion gases. Whatever the theory, igniters for rockets contain an electric fuse and a pyrotechnic composition in a suitable container. On initiation, the igniter produces hot gases, hot particles and radiation, which heat the burning surfaces of the charge to a temperature sufficiently high to initiate combustion. Many propellants do not produce self-sustained combustion below a certain low-pressure threshold. Igniters are supposed to overcome this threshold pressure. Propellant characteristics for the design of an igniter are given in Table 9.1.

Propellant charges are generally fitted with a head-end ignition system, but this adds to undesirable dead weight after combustion. In addition, the igniter casing can damage the propellant bore and structurally it is unsafe to keep the igniter at the head end. In some designs, the head ends are closed and safety prevents placement of the igniter at the head end. A recent trend is to use aft end or nozzle end ignition systems. In this case, the igniter is structurally separated from the motor casing. It is ejected as soon as propellant initiation is completed. This type of ignition system needs a higher igniter charge, as the loss of unburnt igniter charge is higher. This suppresses any ignition peaks and is preferred for staged rockets, where it can be mounted at the top of a jettisoned stage for the initiation of the next stage.

The igniter composition is obtained in various forms like powder, pellet or their combinations. A typical igniter composition consists of a fuel, an oxidizer and a binder. The fuel should have a high heat of combustion, along with physical and chemical stability in the operating temperature range of −40 °C to +60 °C. It should be non-hygroscopic and easily pulverizable. It should be readily oxidized and its combustion products should produce the desired effect. The ingredients must be readily available, stable over the operating temperature range and non-hygroscopic. The binder should reduce sensitivity towards impact and friction and increase the mechanical properties of the produced pellets. It must be compatible with the fuel and the oxidizer and should give sufficient wetting to the igniter ingredients for processing in the granular form. For reprocessibility, it must be thermoplastic in nature.

Igniter compositions typically contain magnesium (Mg) or boron (B) powder with potassium nitrate (KNO_3) in a pressed form. For igniting cordite or plastic propellants, an Mg/KNO_3 based composition, which gives little ignition shock is preferred. In general, for double base propellants, igniter compositions producing hot gases only are preferred while for composite propellant initiation, hot particles in the streams of igniter combustion products are desired. For smaller rockets, pyrotechnic compositions are sufficient to produce high temperature gases at the desired pressure level in the chamber free volume but for large size booster applications, another smaller rocket is used for initiation of the main charge. This system is called a pyrogen ignition system, which is initiated by a pyrotechnic ignition system.

The ignition process has been studied by a number of investigators. Most of early ignition studies were carried out on gem primers. Different propellant compositions ignite differently. In general, the ignition delay is shorter for fast burning double base propellants and longer for slow burning compositions.

9.2   Ignition Theories

A number of attempts have been made to establish a satisfactory ignition theory. The ignition of a solid propellant is a fast, short-lived transient heat transfer phenomena, which starts with triggering an electric pulse and terminates at a point in time matching the establishment of a stable combustion front at the propellant surface. During this short period several events take place simultaneously, such as initiation of igniter composition, creation of hot igniter gases with in-built solid unburnt igniter charges, filling of rocket chamber free volume by igniter gases, heating of propellant surface to self-ignition temperature, establishment of flame front at the propellant surface, flame spreading and establishment of stable self-sustained burning front at the propellant surfaces. Although these events do not occur in tandem, a scheme to predict ignition delay, ignition pressure peaks, kinks and other abnormalities require theoretical studies. Overall, it is a heat transfer phenomenon but the role of fluid mechanics, phase changes, computational fluid dynamics, and chemical kinetics cannot be neglected.

Several theories have been proposed for the ignition of solid propellants like solid phase thermal theory, hypergolic (self-ignition at room temperature), heterogeneous ignition and gas phase ignition theory. As per the solid phase thermal theory, the exothermic reaction established at the surface of the propellant is heterogeneous in nature and is primarily a solid-state phenomenon, while the gas phase theory proposes that the initiation reaction takes place at the vapor boundary level. As per two stage ignition models, the first stage deals with radiation flux in which vaporization of the condensed phase by surface decomposition is established. As radiation flux decreases, the flame approaches the surface and becomes dominant, relegating the condensed phase decomposition to a secondary level.

Practically all observed facts are well explained by the proposed theories. The ignition delay is shortened when the total pressure is increased by a non-oxidizing neutral gas. The ignition delay is also reduced, if the pressurization rate is increased. It is also observed that a maximum temperature and heat is produced at stoichiometric fuel concentrations.

Theoretical studies on the combustion of the Mg–Teflon–Viton (MTV) pyrotechnic composition have been carried out using the NASA Lewis 76 computer model. According to this, a maximum temperature and heat are produced at stoichiometric fuel concentrations and the reaction products are changed significantly with different concentrations of Mg. Teflon and Viton have only a minor effect on the thermodynamic system when compared to Mg. The major reaction products are MgF_2, Mg and solid carbon. MTV decomposes and ignites at a higher temperature than the $B-KNO_3$ system. For a fuel-rich propellant system, MTV compositions are preferred. Smokeless igniters consisting of NC/NG and 10% nickel powder have been developed. The addition of Ni increases the flame temperature and gives improved combustion behavior at low pressure.

9.3 Igniter Design

Igniter design for rockets has been more of an art than a science. It has been a practice in the development of a new rockets that space requirements for propellant, grain support, nozzle, fins and grid are carefully designed and then whatever space is left over is used for the igniters. The igniter design is influenced by the size and shape of propellant charge, its composition, location of igniter in the rocket chamber and the amount of free volume inside the chamber.

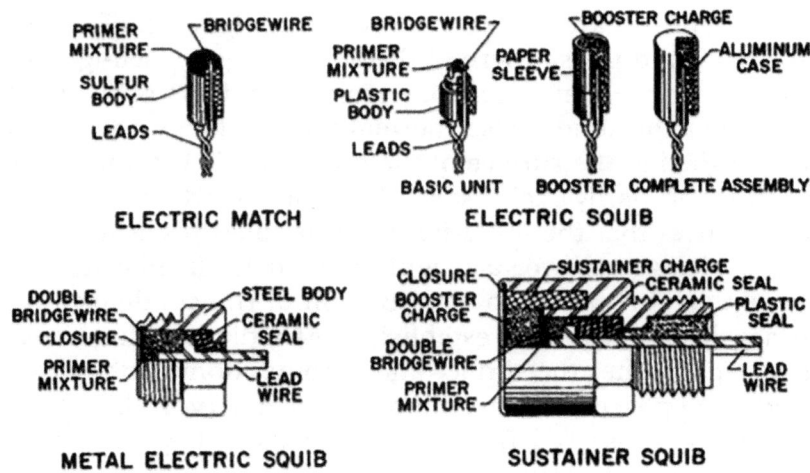

Figure 9.1   Typical electric initiators for solid propellant igniters.

A heat sensitive material is normally used as a bead to the bridge wire. The priming mixture for the squib contains lead azide/lead styphnate. For special requirements, however, complex mixtures are prepared, which may contain lead azide, $KClO_4$, Sb_2S_3, and a synthetic resin as the binder. Figure 9.1 gives details of the different electric initiators of a typical pyrotechnic igniter.

Metal oxidants have replaced black powder in new ignition systems. The most common mixtures contain Mg/Al and KNO_3 or $KClO_4$, with binders. The mixture is pressed into pellets. A variety of sizes and shapes of pellets are used. Igniter containers in the early days were made of cotton cloth. However, due to the hygroscopicity of black powder and other igniter compositions, containers suited for hermetic sealing have been used. The plastic containers, used in the beginning, were relatively inexpensive and could be sealed and made in any desired shape or size. However, most of the materials were brittle and were susceptible to breakage during rough handling. Moreover, these containers absorb NG in double base propellants. Tin containers were used for artillery rockets. This type of container works by a violent rupture of the container or by blowing off of the cover. However, reproducibility depends upon the quality of crimping and gauges of metal. Subsequently, igniter containers made of aluminium or steel were used.

Ignition devices generally follow the contours of the containers. However, their shapes are controlled by their location in the rocket chamber. For most of the rockets, it is advantageous to locate the igniter at the head end. The reaction products of igniters are expected to sweep over the entire length of the propellant charge. However, some rockets have nozzle end igniters and a few others have internal igniters. Fin-stabilized rockets generally have a large length to diameter ratio and hence there is no problem in utilizing the internal space at the head end for the igniter. In large rockets, the igniter can be located at the center of the head closure or it could be in the form of a torus or ring. Large units have several squibs connected in parallel to ensure the effectiveness of squibs in initiating large quantity of igniter material. For large size case bonded motors generally pyrogen igniters (propellant compositions) are used. Spin-stabilized rockets have severe limitations on overall length. In this type of system, any increase in length to accommodate the igniter is unwarranted. Hence, the igniter is usually placed at the nozzle end. By building the igniter into the nozzle plate, even the need for a container is eliminated. Many times because of constraint of configuration of rocket propellant charge, it becomes necessary to locate the igniter in the interior of the propellant charge. Figure 9.2 shows a few typical igniters for solid propellants.

9.4 Igniter Qualification

The qualification of igniter composition is carried out using closed vessel (CV) firing and determination of the calorific value and sensitivity. The ballistic properties of the igniter are ascertained

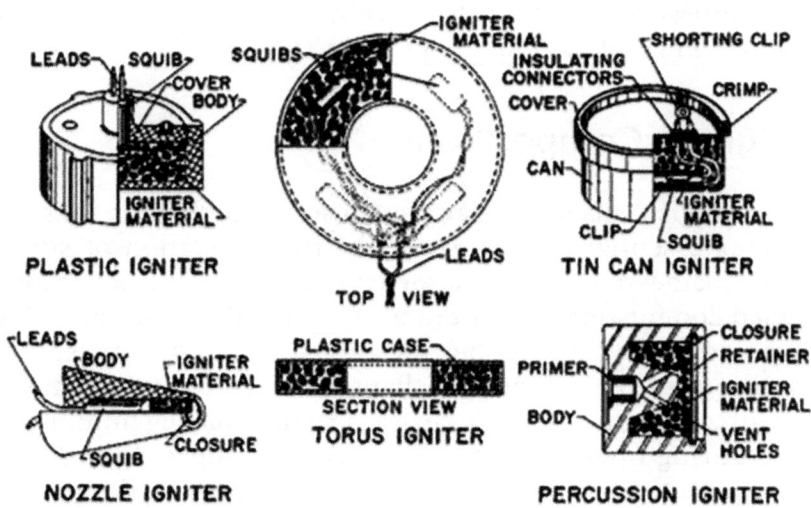

Figure 9.2 The different types of igniters for solid propellant rockets.

Table 9.2 Properties of typical igniter compositions.[a,b]

No.	Type of igniter	Calorific value (cal g^{-1})	Solid:gas	Application
1	Gunpowder	730	60:40	Double base propellants
2	Mg/KNO$_3$/PEC	2100	97:3	Composite propellants
3	B/KNO$_3$/PEC	1700	97:3	CMDB and high-altitude rockets
4	Al/KClO$_4$	2550	93:7	Composite propellants
5	Al/NH$_4$ClO$_4$	2570	80:20	Composite propellants
6	Mg/PTFE/Viton (MTV)	2200	93:7	High energy and fuel-rich propellants

[a]PEC: plasticized ethylcellulose.
[b]PTFE: polytetrafluoroethylene (Teflon).

generally by evaluating it in CV at a low loading density (~0.01 g cm^{-3}). The igniter composition, placed in a cotton bag, is initiated by a suitable squib. The combustion pressure, ignition delay and time to reach maximum pressure is measured as a function of time. The amount of heat evolved by the combustion of pyrotechnic composition is experimentally determined by burning the composition in a bomb calorimeter and is theoretically calculated using Hess's Law. For experimental determination, a certain weighed amount of composition is burnt in a bomb calorimeter, placed in a water bath and the calorific value is determined by the rise in temperature of the water. The Bruceton staircase method and Julius Peter apparatus determines the impact and friction sensitivities, respectively. The properties of some of typical igniter compositions are given in Table 9.2.

9.5   Igniter Compositions

Igniter compositions for the ignition of solid propellants are selected on various essential requirements like the generation of sufficient pressure inside the rocket motor chamber for a steady-state self-sustained combustion wave front at the propellant surface; bringing any exposed uninhibited propellant surface (or a part of it) to the auto-ignition temperature of the propellant; being able to withstand rapid depressurization due to nozzle closure ejection; having minimal interference during operation of the rocket after successful initiation; distributing the combustion gases of the igniters to the free space inside the rocket without any impingement or thermal shocks *etc.* Many of these requirements are dependent on design of the igniter canisters, selection of canister material, distribution of igniter holes, location of igniters and similar interface features. However, compositions for igniters are important contributors for successful solid propellant ignition. Although passing remarks for different igniter compositions were made in earlier sections of this chapter, some of the igniter compositions can pave the way for future developments.

Gun powder, a mixture of charcoal, potassium nitrate and sulfur has been one of the early igniters for the initiation of solid propellants and it still continues to hold the position as a probable candidate for rocket propellant igniters. It has a low heating value and has a high gaseous content, but the ease of manufacturing, availability of raw materials and wide range of performance parameters makes it, probably, the best choice for ignition of all classes of propellants. Although doubts are raised about its performance at high altitude and in a moist atmosphere, but use in larger rockets offsets such doubts. Now, it is used for the ignition of double base propellants, composite propellants and CMDB propellants. Incidentally, it is basically a composite propellant with the capability to explode in certain confinement conditions.

The most common type of igniters for solid propellants is a mixture of solid metallic fuel and a solid oxidizer, pressed as pellets. Amongst metallic fuels, boron (B), magnesium (Mg), aluminium (Al) *etc.* are used. Nitrates (NO_3^-), perchlorates (ClO_4^-) *etc.* are the main solid oxidizers. They are granulated and mixed together before pressing into pellets. The most popular igniter composition is the B–KNO_3 based composition. B–KNO_3 has easy ignition at very low pressures, and produces low sensitivity of burning rate to pressure. The burning rate of the order of 42–45 mm s^{-1} with a pressure index of 0.35 is possible by this composition. The heating value of this composition is around 1500–1700 cal g^{-1}. Al/$KClO_4$ has a high energy content but is difficult to ignite at low pressures. It has a lower burning rate of 10 mm s^{-1} with a burning rate exponent of around 1 (unstable burning). A high energy content and near unity burning rate exponents are favorable features of this composition.

Sometimes, Teflon is employed in place of an oxidizer for better bonding and structural integrity of pellets. Mg–Teflon pellets have energy higher than conventional B–KNO_3 compositions, but the burning rate (10 mm s^{-1}) and burning rate exponent (0.22) are quite low. Mg-Teflon is generally characterized by very low pressure burning-rate exponent and low gas content, has energy output strong in the infrared region and the energy content is approximately equivalent to that of B–KNO_3. A further extension of this composition is MTV (Magnesium,

Teflon, Viton), which has been employed in rocket propellant ignition. For finalizing the MTV compositions for a given rocket application, the magnesium/Teflon ratio is kept constant and the Viton content of the mixture is increased to investigate the effect of binder content on the heat of explosion. This exercise results in an increase of the heat of explosion with an increasing Viton percentage because the energy producing magnesium content effectively goes down toward the stoichiometric value. Theoretical studies have shown that the heat of explosion of magnesium rich compositions decreases with the increasing fractions of both magnesium and Viton in the mixture. The reason for the latter case is that the replacement of Teflon by Viton increases the carbon fraction in the mixture and introduces small amounts of HF and MgH as combustion products. In other words, the effect of magnesium content on the heat of explosion is dominant up to a certain Viton fraction (12%). However, after this Viton fraction the heat of explosion starts to decrease with increasing Viton content. This implies that the heat of explosion is predominantly affected by the increasing Viton content beyond the value of 12%.

In general, MTV-based pyrotechnic compositions possess many properties suitable for their use as rocket motor igniter materials. In particular, they have very good igniter material characteristics, due to the existence of hot solid and liquid particles and reactive condensable species in their combustion products. The existence of hot and reactive particles enables the fast ignition of the solid propellant surface by almost all possible modes of heat transfer. Thus, the MTV-based pyrotechnic compositions have very high ignition effectiveness. MTV mixtures have also been used for base bleed systems, which are normally very difficult to ignite. Most of the studies on MTV compositions are about their ballistic properties, which are of great importance in designing igniters for new solid rocket propellants. It has been shown that the burning rate of the MTV igniter depends on the porosity of the composition, its charge length, and the size of its particulate components. The size of the magnesium particles used in MTV formulations affects not only the burning rate, but also the sensitivity of the igniter. The influence of the density, reduced pressure (to simulate high altitude conditions) and high rotation speeds (to simulate spinning projectiles) on the burning rate of the MTV composition has been well investigated, and the relationship between the combustion and sensitivity characteristics has been established. Another important characteristic of the MTV mixture is its aging caused mainly by the reaction of magnesium with water. An accelerated aging study by using IR, X-ray, and bomb calorimeter has shown that the energy and mechanical properties of MTV igniters are deteriorated during the aging depending on the temperature and relative humidity of the air, and the binder system provides only partial protection against aging.

The replacement of magnesium by other metals (aluminium, titanium, boron *etc.*) in MTV composition has also been attempted. Aluminium-based compositions give higher flame temperatures, but the biggest defect of these mixtures is their initiation. These mixtures would be very perspective in terms of application if their initiation problem is solved. The heat conductivity for titanium (Ti) is 10 times smaller than the heat conductivity for magnesium and this defines the characteristics of MTV mixtures, containing titanium instead of magnesium. The maximum burning pressure, burning rate and flame temperature are lower than for mixtures with aluminum and magnesium, and their changes with added content of titanium in the mixture are similar. The mixtures with the content of titanium below 48% could not be initiated. Similar action occurs with boron in MTV. The small heat conductivity of boron (100 times smaller than magnesium), makes the igniter composition almost impossible to ignite with an electric squib of any type. If they are initiated while in a loose condition, big flames burn across the surface with the tendency to extinguish, because the heat directing towards the non-reactive mixture is very weak. The burning of these mixtures while in a compressed condition is much more stable because the burning of the non-reactive mixture is helped by burning particles. So, such replacements are not very helpful in igniter applications.

The Combustion Mechanism of Solid Rocket Propellants

10.1 Introduction

For effective propulsion, propellants undergo combustion reactions and are transformed into gaseous products. The mechanism by which combustion takes place differs for each class of propellant. In the case of liquid propellants, droplet formation, vaporization and flow rates are important factors, whereas for solid propellants, propellant composition, surface conditions, pressure condition and a host of other parameters control combustion behavior.

Since solid propellants contain various materials like oxidizers, polymers, nitro compounds, metallic powders and burn rate catalysts, their burning process depends on the physical and chemical properties of these materials. In general, a combustion wave of solid propellants consists of successive reaction zones: the subsurface and surface reaction zones, the diffusion mixing zone between gaseous oxidizer and fuel fragments and the high temperature luminous flame zone. The thermo–chemical properties of propellant ingredients are major parameters that alter the burning rate characteristics. The diffusion mixing process also plays an important role in determining the burn rate characteristics.

The combustion mechanism of both double base and composite propellants has been studied extensively. The burning rate of AP-polymer

Solid Rocket Propellants: Science and Technology Challenges
By Haridwar Singh and Himanshu Shekhar
© Haridwar Singh and Himanshu Shekhar 2017
Published by the Royal Society of Chemistry, www.rsc.org

based propellants is increased by the addition of iron oxide or organic iron compounds (ferrocene derivatives) and is decreased by the addition of lithium fluoride. The addition of lead/copper salts alters the combustion mode of double base propellants, resulting in super rate burning and a plateau, where the pressure index value is very low or zero.

The combustion process of a propellant is dependent on pressure, propellant temperature, chemical composition and their properties. Figure 10.1 shows the flame structure of propellants. In zone I (solid phase), no chemical reaction occurs and the temperature increases exponentially from the initial propellant temperature (T_o) to the decomposition temperature (T_d) by solid phase heat conduction. In zone II, the temperature increases relatively slowly from T_d to the burning surface temperature T_s. Zone II is a thin layer, where phase changes from solid-to-liquid and liquid-to-gas take place, producing gaseous species at the burning surface. In zone III, *i.e.* gas phase, the temperature increases rapidly from T_s to T_f (flame temperature). Since the thickness of the combustion wave is very thin, measurements of the gaseous species produced in the combustion wave by means of gas sampling and optical techniques becomes difficult. With the help of fine thermo couple, measurement of temperature distribution is possible.

Since the reaction rate in the gas phase decreases with decreasing pressure, the thickness of the reaction zone increases with lowering the pressure. Hence, a detailed measurement of the flame structure is

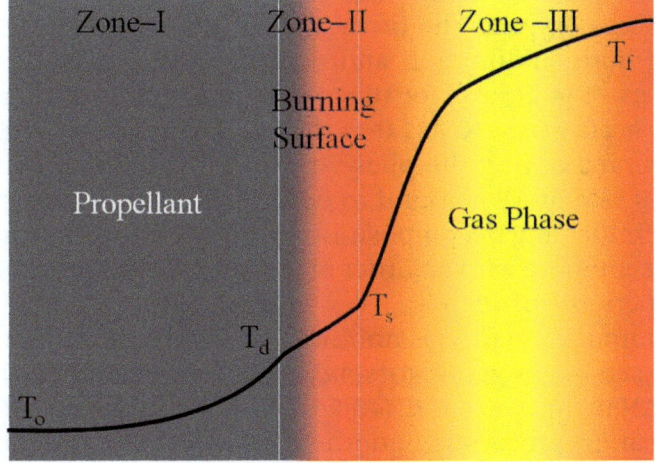

Figure 10.1 The combustion wave structure of propellants.

possible only at low pressures. The reactive species produced in zone II are ejected from the burning surface and react to form a luminous flame in zone III.

10.2   Thermal Analysis of Crystalline Ingredients

Composite propellants are a combination of crystalline oxidizers and polymeric binders and their combustion behavior is dependent mainly on the combustion behavior of individual ingredients. Similarly, double base propellants have two main molecules (both containing fuel and oxidizer), and the combustion wave structure is a function of their individual combustion. To get an insight into the combustion wave structure, thermal analysis is carried out at a pre-specified heating rate. Differential thermal analysis (DTA) is one tool for such a study, where a sample is heated against an inert sample and the difference in heat generation or heat absorption at different temperatures between sample and reference is obtained. This gives a heat of reaction for the different transitions in the form of exothermic or endothermic behavior. Another tool is thermogravimetric (TG) analysis, where mass loss or gain of sample at a different heating rate is monitored.

Ammonium perchlorate (AP) is one of the important ingredients of composite propellants. The DTA of AP at a low heating rate (0.33 K s^{-1}) reveals that at 520 K, an endothermic peak occurs, which corresponds to a crystal phase transition of AP from orthorhombic to cubic structure. The heat of reaction of these changes is -85 kJ kg^{-1}. There is no mass loss during this process. An exothermic reaction occurs between 607 K and 720 K, where gasification of AP takes place with the generation of excess oxygen. The process is quite complex in this phase, containing both sublimation and melting. The activation energy is calculated as $+134 \text{ kJ kg}^{-1}$, for this decomposition. The melting of AP is stated to occur at 833 K. As far as the TG of AP is concerned, no mass loss is observed during the phase transition. However, a single step major mass loss of AP is found to occur during the exothermic reaction phase indicating decomposition. In composite propellants, AP is stated to decompose with an order of reaction of two, and an adiabatic flame temperature of 1205 K.

Ammonium nitrate (AN) is another significant propellant oxidizer, generally used in gas generators, because of production of low burning rates. AN melts at around 443 K and begins to gasify above 480 K. The gasification process is endothermic at lower temperatures and becomes exothermic at high temperatures. This change of decomposition behavior is represented by the following reactions.

$$\text{Endothermic: } NH_4NO_3 \leftrightharpoons NH_3 + HNO_3$$
$$\text{Exothermic: } NH_4NO_3 \leftrightharpoons N_2O + 2H_2O$$
$$\text{Overall: } NH_4NO_3 \leftrightharpoons N_2 + 2H_2O + 1/2O_2$$

The overall reaction shown above is highly exothermic and production of oxygen molecules is also ensured. This occurs at high temperatures. It can be concluded that due to the endothermic reaction at low temperatures, it is difficult to ignite it, but as the temperature rises, the conversion of endothermic into exothermic, results in high flammability and even decomposition is possible above 600 K.

Nitramines (RDX and HMX) are energetic ingredients used in both composite and double base propellants. The melting temperatures of RDX and HMX are 477 K and 553 K, respectively. HMX has been used in propellant formulations and in TG it shows single stage decomposition starting at around 550 K. It is completely consumed at 553 K. In DTA, first endothermic peak at 462 K represents crystal phase transition from β to α form and this is not associated with any mass loss (no trace in TG). The second endotherm occurs at 550 K, which corresponds to melting. The final exotherm at 553 K represents a gas phase reaction.

The thermal decomposition of ammonium dinitramide (ADN) has melting at 328 K, as observed in DTA. The onset of decomposition is at 421 K and exothermic peak occurs at 457 K. The decomposition of ADN is initiated by formation of ammonium and hydrogen dinitramide. Hydrogen dintramide decomposes into AN and N_2O. The final decomposition products after gas phase reaction is O_2, H_2O and N_2. The adiabatic flame temperature may reach 3640 K. The combustion wave of ADN has three distinct layers:

(i) A melt layer, where temperature is more or less constant.
(ii) A preparation zone, where the temperature rises to around 1300 K.
(iii) A flame zone, which is at a certain stand-off from the melt layer and has the final combustion products.

Hydrazinium nitroformate (HNF) melts at 397 K and decomposes at a relatively lower temperature (439 K) than AP or ADN. HNF decomposes in two exothermic stages:

(i) First stage: 60% mass loss occurs between 389–409 K.
(ii) Second stage: 30% mass loss occurs between 409–439 K.

The combustion wave structure of HNF does not have any melt layer and it is predominantly containing two gas phases—one at the surface and another at a certain stand-off from the burning surface.

10.3   Combustion Mechanism of Composite Propellants

With regards to the flame structure of AP based composite propellants, AP particles mixed with a polymeric fuel binder and decompose to produce ammonia (NH_3) and perchloric acid ($HClO_4$) at the burning surface of the propellant and react to produce mono-propellant flamelets above the burning surface. The fuel fragments produced by the decomposition of the polymeric fuel react with the AP mono-propellant flamelets to produce diffusion flame streams above the burning surface. Thus, the flame structure of AP composite propellants is dependent on the physical and chemical properties of ingredients mixed within propellants, such as concentration of binder, size of AP and presence of burning rate catalysts. The burning rate of propellant increases linearly with pressure in $\log P$ vs. $\log r$ ($r = aP_c^n$). The temperature increases smoothly in the solid phase from T_0 to T_s and increases rapidly from T_s to T_f. The decomposed gases from AP and binder diffuse into each other and produce diffusion type flamelets above the burning surface. Since the shape of these flamelets changes in space and in time, the gas phase structure becomes highly heterogeneous. The reaction time decreases linearly with increasing pressure.

A number of combustion models have been developed during the last three decades to explain the combustion mechanism of AP based composite propellants. The granular diffusion flame (GDF) model proposed by Summerfield is based on the concept that the fuel and oxidizer gasify at the burning surface, leaving the surface in pockets of gases that diffuse together. Although this model was quite popular, it could not describe the variations in the burn rates observed for varying pressure, oxidizer particle size and solid loading. Generally, fine AP causes increased burn rates, but at low and high pressures this effect is diminished. Likewise, an increase in AP content leads to an increase in burn rates. Thus, the GDF model had limitations. According to the GDF model, the burn rate can be calculated by the equation: $1/r = a/p + b/p^{(1/3)}$ or $p/r = a + bp^{(2/3)}$, where a and b are constants, termed as chemical reaction and diffusion time parameters, respectively. The particle size effect is contained in parameter 'b'. An alternate form of the burn rate and pressure relationship has been proposed by Penner $(1/r)^2 = (a/p)^2 + (b/p^{(1/3)})^2$.

Subsequently, Hermance proposed a model of a heterogeneous reaction at the burning surface between the oxidizer and the binder, creating an increased surface area. The vapor phase description by Hermance considers the essence of heat transfer to the condensed phase without attempting model details. The vapor phase flame gets closer to the surface as the size of AP is decreased, leading to a higher heat transfer, resulting in increased burn rates.

In the early 1970s, Beckstead *et al.* published their multiple flame model called the Beckstead, Derr and Price (BDP) model. According to this model, a complex interaction between oxidizer mono-propellant flame and two different diffusion flames takes place above the oxidizer–binder interface. Figure 10.2 illustrates this model. The physical picture presented was comprehensive to explain most of the observed facts satisfactorily. According to the BDP model, the burning rate is dependent highly on particle size due to the primary diffusion flame. For very small particles, the diffusion aspect of the primary diffusion flame can be reduced and the kinetic aspect becomes dominant. However, for large particles, a monopropellant flame dominates and reduces the particle size effect. The main recommendation was that the primary diffusion flame was the dominating factor in determining burn rates of AP composite propellants and at high pressures. For large particles of AP, a mono-propellant flame becomes dominant.

AP and HMX have similar monopropellant combustion characteristics but when mixed with a binder, the resultant burn rates vary by

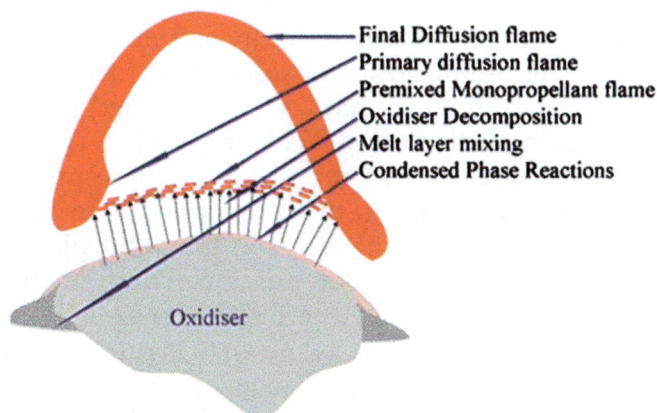

Final Diffusion flame
Primary diffusion flame
Premixed Monopropellant flame
Oxidiser Decomposition
Melt layer mixing
Condensed Phase Reactions

Oxidiser

Figure 10.2   A schematic burning pattern for a composite propellant by the BDP model.

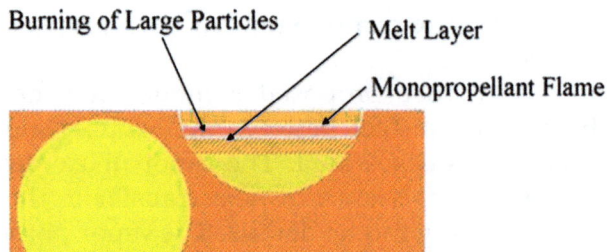

Figure 10.3 The combustion mechanism for the HMX composite propellant.

as much as an order of magnitude. The burn rates of HMX-based propellants are much lower than those of AP-based propellants and show much smaller particle size dependence (Figure 10.3).

The AP diffusion flame is much more energetic than the AP monopropellant flame and results in higher burn rates. HMX composite propellants have lower T_f, employing a less energetic diffusion flame than the monopropellant flame of HMX, leading to reduced burn rates. It was concluded that the dominant mechanism in propellant combustion is related to the primary diffusion flame when compared to the monopropellant flame.

10.4 The Combustion Mechanism of Double Base Propellants

Since double base propellants are homogenous, their combustion flame structure appears to be homogenous and one dimensional along the burning direction. The decomposed gases produced at the burning surface are premixed with oxidizer and fuel on a molecular scale. Four combustion zones have been identified for DBP: foam, fizz, dark and luminous flame zones (Figure 10.4).

There exists a heat affected zone in double base propellants. No chemical reaction takes place in this zone, but due to heat feedback from the flame side, the temperature in the solid phase increases to the onset temperature of a solid-phase reaction. At the surface a solid phase reaction zone exists, where nitrogen oxide and aldehydes (RCHO:HCHO *i.e.* formaldehyde, as shown in Figure 10.4) are formed. The reaction is endothermic, but subsequent reaction between nitrogen oxides and aldehydes is exothermic. This makes the overall reaction in the solid phase exothermic. This solid-phase reaction zone has popping of gaseous form through melt layers and this zone is also

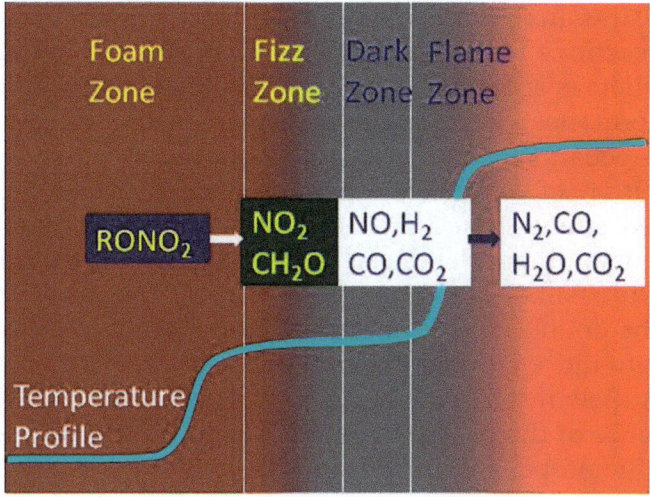

Figure 10.4 The combustion wave of double base propellants.

called foam zone. The reaction in the fizz zone is very rapid and the formation of NO, CO, H_2O, H_2 and carbonaceous matter occurs in this zone. The dark zone is basically a cushion (preparation zone), where the oxidation of reaction products in the fizz zone occurs. The rates of reaction are slow, except when the pressure and temperature is high. This zone acts as an induction zone for further rapid reaction and heat release in the flame zone. However, at low pressure, no flame zone is produced, because of a too slow conversion of NO to N_2. The temperature profile of double base propellants has two steps: one occurring in the solid phase reaction zone to bring the temperature from room (or soaking) temperature to onset/decomposition temperature and next is to get the final flame temperature. Since these zones have a very small thicknesses, the ramped rise in temperature appears as step-rise.

The thickness of the fizz zone is a function of the chemical kinetics of the reactions occurring at the propellant surface. This is dependent on pressure. As the pressure increases, the thickness of this zone reduces making the ramp rise in temperature steeper. This steep rise in temperature indirectly points towards an increased heat transfer from the flame zone towards the solid propellants. The same occurs for the dark zone. At low pressure, below 35 kg cm^{-2} (ksc), the luminous flame zone is not dominant and heat transfer from the dark zone to the fizz zone controls the reactions. As the pressure increases to 70 ksc, the flame zone comes very close to the burning surface and heat transfer from the gas phase to the burning surface controls the burn rate. Thus, in the case of double base propellants, the dark zone/ fizz zone reactions play an important role in controlling the burning process, whereas at higher pressures gas phase reactions assume a higher importance. The combustion mechanisms of catalyzed/platonised propellants are discussed in Chapter 12.

The addition of AP in a double base matrix results in composite-modified double base (CMDB) propellants. AP introduces small independent luminous flamelets (as discussed earlier) in the flame. As the size of the flamelets is bigger than the thicknesses of the various zones of the double base propellants' flame structure, the flamelets extends from the solid phase reaction zone to the flame zone of double base propellants. The ultimate result of these AP flamelets is the disappearance of the dark zone and a merger of the luminous flame zone with the propellant burning surface. Although the introduction of AP makes the flame structure heterogeneous, the heat feedback becomes superior to the double base propellants. AP releases oxygen and the double base propellant flame is oxygen-deficient. So the net effect of adding AP, is a shift of chemical reaction towards stoichiometric ratio. This increases both the reaction rate and flame temperature. Introduction of AP results in formation of diffusion flame at the propellant surface.

10.5 Combustion Mechanism of Nitramine-Based Propellants

Nitramine (RDX/HMX) based solid propellants along with energetic binders (GAP), energetic plasticizers (Bu-NENA) and high energy additives (CL-20) have great potential to offer very high energy (Isp ~ 300 s) with plume signature advantage, over the state-of-the-art composite propellants based on AP/HTPB/Al. There has been considerable research over the years attempting to elucidate the combustion mechanism of nitramines and nitramine-based propellants. However, the mechanisms proposed so far are valid under limited temperature and pressure conditions. A considerable controversy and uncertainty still exists regarding the initial pathways. The major problem encountered is the short life of highly reactive decomposition intermediates. It becomes difficult to experimentally observe in a quantitative way many of the decomposition species. Part of difficulty lies in the lack of reliable data on potential decomposition products. A quantum chemistry approach has been applied by researchers to calculate the heat of formation and free energies of nitramines and double base matrix.

Since DBP are fuel rich, no exothermic chemical reactions are possible between their combustion products. The flame temperature and Isp of nitramine-based CMDB propellants are therefore approximately equal to the weight average value of each component. The crystalline nitramine particles mixed with NC/NG of DBP melt, decompose and gasify at the burning surface of the propellant. Different process occur before the nitramine particles produce their own mono-propellant flamelets above the burning surface. The burn rate of nitramine based CMDB propellants decreases with an increase in concentration of nitramines, despite an increase in the energy content per unit mass of propellant.

It has been reported that the mole fraction of NO_2 produced by the initial decomposition of HMX is less than that of NO_2 produced by decomposition of DBP matrix. Thus, the ratio of NO_2/aldehyde of DBP is reduced by the addition of HMX. This indicates that HMX shifts the equivalent ratio of NO_2/aldehyde towards fuel richness, thereby decreasing the reaction rate in the fizz zone and reducing the heat feedback from the gas phase to the burning surface. Consequently, the burn rate of HMX-based CMDB propellants decreases when compared to DBP. It has been suggested that diffusion controlled effects are not introduced by the inclusion of HMX, as happens in the case of AP. Kubota has reported that the flame structure of the base matrix is not altered by the addition of HMX. It has been found by other researchers that HMX does not ignite at the burning surface of the propellant. However, it ignites and burns in the gas stream above the propellant burning surfaces. They argued that HMX could be considered as a diluent. Since HMX decomposition occurs at a longer distance from the burning surface, the heat feedback to the surface is not the rate controlling process. Another possible explanation could be the physical change of the burning surface from the melt layer to the cratered surface and the delayed ignition of nitramine crystals.

The gas phase flame chemistry of the RDX flame has been studied theoretically, using a steady laminar one dimensional premixed flame model. As per a new description for the chemistry of the nitramine flame, there exists a primary flame zone composed of an endothermic pyrolysis region, followed by a combustion zone. However, the primary flame maintains a spatial separation from the luminous flame zone. It is not the primary flame chemistry near the surface but rather the secondary luminous flame chemistry that couples with the overall heat and mass transport to control the burn velocity. Another view in this regard suggests that in the condensed phase combustion, evaporation and thermal decomposition of nitramines occur simultaneously with negative heat release below 5–10 atmospheres. However, the heat release becomes positive beyond the threshold of low pressure. The gasification law and heat release law do not change in the solid propellants, but the dark zone laws and dependence on temperature sensitivity changes significantly with the inclusion of HMX. In fact, studies on the chemical pathways at the propellant burning surface revealed that while high burn rate materials such as GAP release a large amount of energy in the early reaction steps at the propellant surface, the intermediate burn rate materials like RDX/HMX decompose initially by thermally neutral sets of reactions. High heat release is delayed and divided between a heterogeneous condensed phase.

Ballistic modification of nitramine-based CMDB propellants is very much sensitive to the method of processing the propellant. The burn rate enhancement and reduction of 'n' values is highly dependent on the degree of gelatinization of NC in the double base matrix of CMDB propellants. A minimum of 35–40 passes through hot rollers are needed to obtain reproducible ballistic properties of ballistically modified extruded CMDB propellants. The comparatively lower catalytic/plateau effect of CMDB propellants than DBP with similar lead/copper salts may be partly due to non-uniform distribution of ballistic modifiers in the spheroidal dense nitrocellulose.

Zarko has reported that RDX/HMX burns without appreciable heat release in the condensed phase and an endothermic vaporization dominated on the surface. The up-to-date models developed for nitramines takes account of the global reactions in the liquid phase, detailed reaction kinetics in the gas phase and non-equilibrium evaporation at the propellant surface. It was found that the larger the difference between MP and burning surface temperature, the narrower the combustion stability domain. Preliminary calculations show that the radiation driven combustion of evaporated nitramine may occur in the gasification regime, which means almost negligible heat feedback from the gas phase.

From the account given above, it appears that the flame structure of RDX/HMX-CMDB propellants is similar to DBP, consisting of various reaction zones: foam, fizz, dark and luminous flame zones, but is different from AP-CMDB propellants. Recently, it has been proposed that there are three parallel reactions occurring in the condensed phase of double base propellants, a first order reaction with NO_2 and second order reaction of complex aldehyde species. The gas phase kinetics have been summarized with four reactions, a first order NO_2 reaction, a second order reaction of aldehydes, a fist order NO–carbon reaction and a second order NO reduction. The agreement between experimental results and numerical results is satisfactory.

Based on quantum chemical calculations, it is suggested that the initial step in the decomposition of nitramines and nitrate esters is NO_2 bond scission. In the condensed phase, energetic radicals formed from initial NO_2 bond scission can be trapped by cage effects. These trapped radicals can be converted to unsaturated non-radical intermediates before the radical can unzip. It is proposed that trapped HONO can add back into the energetic molecule. This produces oxidation products in the condensed phase that normally would not be produced until much later in the flame. This prompt oxidation mechanism is a general feature of both nitramines and nitrate esters. The luminous flame appears down stream in the gas phase of GAP/nitramine propellants and is caused by the exothermic gas phase reaction of nitramines.

10.6 Flame Structure of Modern Solid Propellants Containing HMX and GAP

Glycidyl azide polymer (GAP) is a binder used with advanced solid rocket propellants. GAP is known to have self-sustaining burning characteristics. The decomposition and gasification reaction of GAP occurs in the temperature range of 475–537 K. The gasification can be divided into two exothermic steps: the first step occurs at low temperature, but rapidly; the second step occurs at a relatively slow pace. The low temperature rapid reaction indicates better ignition and sustenance of the combustion reaction.

Although the Isp of nitramine propellants is increased by the use of azide polymers, the burning rate of azide/nitramine propellants is still low when compared to conventional NC/NG based propellants. Lead/copper salts effective as burning rate catalysts for double base propellants are also effective with nitramine/GAP propellants. Typical nitramine/GAP based propellants contain 80% HMX and 20% GAP. Kubota found that the burn rates of HMX/GAP propellants was increased considerably with lead citrate and 0.6% carbon black. The heat released at the burning surface was increased by the addition of catalyst and hence the first stage reaction zone is responsible for the burn rate increase. However, the effect of catalyst diminishes as the pressure is increased. GAP without HMX burns relatively faster with a luminous flame and the flame contains carbonaceous fragments. HMX burns with a luminous flame, which is produced just above the burning surface. However, the mixture of GAP and HMX produces a luminous flame some distance above the burning surface.

10.7 Combustion Characteristics of Advanced Propellants

Looking at the various mechanistic aspects of AP- and HMX-based composite propellants and DBP, it is quite apparent that the primary diffusion flame is the dominant mechanism in AP-based propellants. In contrast, HMX-based propellants are devoid of an effective diffusion flame due to a lack of reactive chlorine species. None of the advanced propellants containing ADN, HNF, CL-20 have chlorine atoms. So by logic, the dominance of the diffusion flame in the combustion of advanced propellants seems to be less important. Thus, the combustion of advanced propellants may be similar to HMX combustion behavior. Before analyzing combustion behavior, a quick look at their thermal decomposition behavior can give an insight into their burning patterns.

The implied meaning of the absence of chlorine is that the particle size effect of advanced propellants in all probability should be small. This is supported by reported observations that for AND-based propellants; larger particles burn faster than those with smaller particles. This indicates that ADN burn very fast, leaving HTPB virtually untouched. It appears that the burning behavior of advanced propellants will depend largely on the relative rate of monopropellant ingredients. Coarse particles would be expected to burn at the rate of monopropellant, whereas fine particles would burn with premixed or diffusional flame based on the properties of mixed ingredients. Thus, the relative monopropellant burn rates of propellant ingredients will be an important factor in controlling the burn rate behavior. However, a high burn rate binder could control the propellant burn rate. The size and concentration of fine particles will have an effect. The propensity for melting the binder can also affect the burn rate, particularly at low pressure, where the melt layer thickness is greater.

Bis-azido methyl oxetane (BAMO) is a solid binder used in advanced solid rocket propellants. It has a complete exothermic gasification at around 500 K with a mass loss of around 35%. Scission of two azidebonds during this process results in the release of nitrogen gas and heat. The remaining part of BAMO gasifies without any exotherms at higher temperatures. BAMO-based propellants could be expected to have burn rates approaching additive monopropellant burn rates due to the very low rate of BAMO. Hence, the net burn rate will be related to the monopropellant burn rate. Monopropellant burn rates of new oxidizers are in the order ADN > CL-20 > RDX. Hence, corresponding propellants may show a similar trend. At around 70 kg cm^{-2} pressure the ratio of mono-propellant burning rates is $3:2:1$ for ADN : CL-20 : RDX. One can expect that propellants based on these oxidizers would produce corresponding burn rates.

GAP has a much higher burn rate than RDX. However, ADN and CL-20 have higher burn rates than GAP. So GAP/ADN- or CL-20/GAP-based propellants would have burn rates approaching to ADN or CL-20 monopropellants. RDX/GAP propellants would burn at the GAP monopropellant rate. Since GAP and BAMO have similar energies, same trend is expected with particle sizes and concentrations of oxidizers. These predictions are need to be verified with the help of experimental data.

The Control and Guidance of Missiles

11.1 Introduction

Missiles are unmanned flying objects, supposed to search, detect, acquire, track and attack targets of a movable or stationary nature. For this, they need maneuverability of their flight path during trajectory. This requirement of modern missile systems requires special features in the missile body. These features are called control surfaces. To actuate a typical control surface for a given change in attitude of the missile, another system, called the guidance system is either mounted on missiles or placed at ground locations and remains in proper correspondence with missile. The guidance system is that part of a missile which decides when, and by how much, the control system must change the trajectory of the missile.

11.2 Missile Control

The missile control is considered synonyms of the generic term 'fin', which refers to any aerodynamic surface on a missile. Missile designers, however, are more precise in their naming methodology and generally consider these surfaces to fall into three major categories: canards, wings, and tail fins. The heart of a missile is the body, which is equivalent to the fuselage of an aircraft. Figure 11.1 illustrates a

Solid Rocket Propellants: Science and Technology Challenges
By Haridwar Singh and Himanshu Shekhar
© Haridwar Singh and Himanshu Shekhar 2017
Published by the Royal Society of Chemistry, www.rsc.org

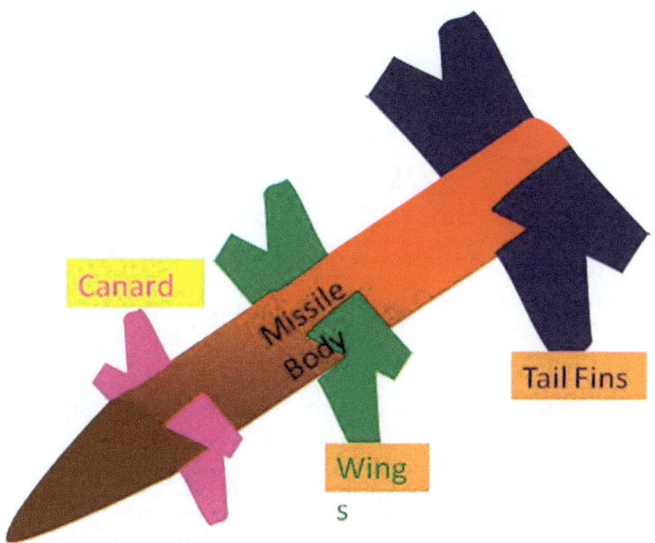

Canard

Missile Body

Tail Fins

Wing

s

Figure 11.1 The nomenclature of missile control surfaces.

generic missile configuration equipped with all three surfaces. Often, the terms canard, wing, and fin are used interchangeably, which can get rather confusing. These surfaces behave in fundamentally different ways, based upon where they are located with respect to the missile's center of gravity. In general, a wing is a relatively large surface that is located near the center of gravity, while a canard is a surface near the missile nose and a tail fin is a surface near the aft end of the missile. The majority of missiles are equipped with at least one set of aerodynamic surfaces, especially tail fins since these surfaces provide stability in flight. The majority of missiles are equipped with a second set of surfaces to provide additional lift or improved control. Very few designs are equipped with all three sets of surfaces.

In order to turn the missile during flight, at least one set of aerodynamic surfaces is designed to rotate about a center pivot point. In so doing, the angle of attack of the fin is changed so that the lift force acting on it changes (Figure 11.2). The changes in the direction and magnitude of the forces acting on the missile cause it to move in a different direction and allow the vehicle to maneuver along its path and guide itself towards its intended target. Canards, wings, and tails (Figure 11.3) are often lumped together and referred to as aerodynamic controls. A more recent development in missile maneuvering systems is called unconventional control. The most unconventional control systems involve some form of thrust vector control (TVC) or jet interaction (JI), which will be explained in Chapter 12.

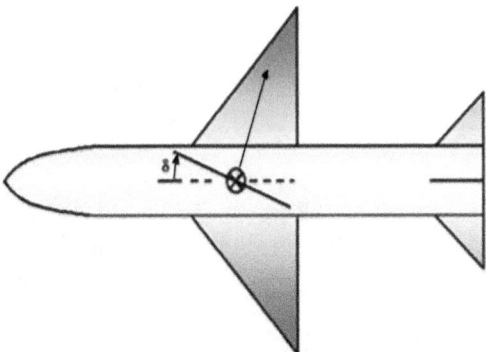

Figure 11.2   The deflection of a control surface on a missile.

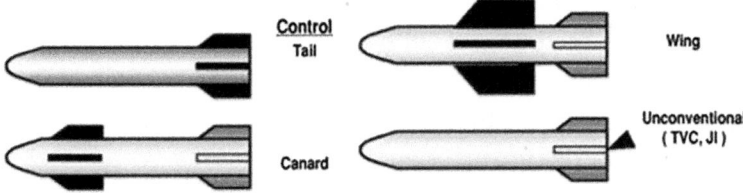

Figure 11.3   Four forms of missile control mechanisms.

11.2.1   Tail Control

Tail control is probably the most commonly used form of missile control, particularly for long range air-to-air missiles like AMRAAM and surface-to-air missiles like Patriot and Roland. The primary reason for this application is that tail control provides excellent maneuverability at the high angles of attack often needed to intercept a highly maneuverable aircraft. Missiles using tail control are also often fitted with a non-movable wing to provide additional lift and improve range. Some examples of such missiles are air-to-ground weapons like Maverick as well as surface-to-surface missiles like Harpoon and Exocet. Tail control missiles rarely use canards, although one such example is the AIM-9X Sidewinder. In addition to missiles, some bombs also use tail control. An example is the JDAM series of GPS-guided bombs.

11.2.2   Canard Control

Canard control is quite commonly used, especially on short range air-to-air missiles like the AIM-9M Sidewinder. The primary advantage of canard control is better maneuverability at low angles of attack, but canards tend to become ineffective at high angles of attack because of flow separation that causes the surfaces to stall. Since canards are ahead of the center of gravity, they cause a destabilizing effect and require large fixed tails to keep the missile stable. These two sets of fins usually provide sufficient lift to make wings unnecessary.

A further subset of canard control of missiles is the split canard. Split canards are a relatively new development that has found application in the latest generation of short range air-to-air missiles like Python 4 and the Russian AA-11. The term split canard refers to the fact that the missile has two sets of canards in close proximity, usually one immediately behind the other. The first canard is fixed while the second set is movable. The advantage of this arrangement is that the first set of canards generates strong, energetic vortices that increase the speed of the airflow over the second set of canards making them more effective. In addition, the vortices delay flow separation and allow the canards to reach higher angles of attack before stalling. This high angle of attack performance gives the missile much greater maneuverability when compared to a missile with single canard control. Many smart bombs also use canard control systems. The most notable of these are laser guided bombs such as the Paveway series.

11.2.3   Wing Control

Wing control was one of the earliest forms of missile control developed, but it is less commonly used these days. Most missiles using wing control are longer-range missiles like Sparrow, Sea Skua, and HARM. The primary advantage of wing control is that the deflections of the wings produce a very fast response with little motion of the body. This feature results in small seeker tracking error and allows the missile to remain locked on target even during large maneuvers. The major disadvantage of the system is that the wings must usually be quite large in order to generate both sufficient lift and control effectiveness, which makes the missiles overall large. In addition, the wings generate strong vortices that may adversely, interact with the tails causing the missile to roll. This behavior is known as induced roll, and if the effect is strong enough, the control system may not be able to compensate.

11.2.4   Unconventional Control

Unconventional control systems are a broad category that includes a number of advanced technologies. Most of the techniques involve some kind of thrust vectoring (Chapter 12). Thrust vectoring is defined as a method of deflecting the missile exhaust to generate a component of thrust in a vertical and/or horizontal direction. This additional force points the nose in a new direction, causing the missile to turn. Another technique that is just starting to be introduced is called reaction jets. Reaction jets are usually small ports in the surface of a missile that create a jet exhaust perpendicular to the vehicle surface and produce an effect similar to thrust vectoring. These techniques are most often applied to high off-boresight air-to-air missiles like AIM-9X Sidewinder and IRIS-T to provide exceptional maneuverability. The greatest advantage of such controls is that they can function at very low speeds or in a vacuum, where there is little or no airflow to act on conventional fins. The primary drawback, however, is that they will not function once the fuel supply is exhausted. Most missiles equipped with unconventional controls do not rely on these controls alone for maneuverability, but only as a supplement to aerodynamic surfaces like canards and tail fins.

11.3   Missile Guidance

Missile guidance deals with ability of a missile to seek out and accurately navigate their way to the correct target without assistance from a human operator. A missile very rarely makes a mistake unless it is misprogrammed by the launching agency. In fact, many of the methods used for missile guidance are the same as those used to navigate manned planes like commercial airliners. Missile guidance concerns the method by which the missile receives its commands to move along a certain path to reach the target.

On some missiles, these commands are generated internally by the missile computer autopilot. On others, the commands are transmitted to the missile by some external source. The missile sensor or seeker is a component within a missile that generates data fed into the missile computer. This data is processed by the computer and used to generate guidance commands. Sensor types commonly used today include infrared, radar, and the global positioning system (GPS). Based on the relative position between the missile and the target at any given point in flight, the computer autopilot sends commands to the control surfaces to adjust the missile's course.

There are many types of guidance used in missiles. Some of which are briefly introduced here.

- Radar command: this system uses two radars, one to track the target and the other to track the missile. The bearing and elevation of the radar trackers is monitored by a computer that directs the control system to alter the missile's trajectory until it reaches the target.

- Radio command: this system uses a human operator to observe the trajectory of the missile and direct its control system using a remote control radio link.
- Wire guidance: this system is broadly similar to radio command, but is less susceptible to electronic counter measures. The command signals are passed along a wire (or wires) dispensed from the missile after launch.
- Inertial guidance: this system is totally contained within the missile and is programmed prior to launch. Three accelerometers, mounted on a platform space-stabilized by gyros, measure accelerations along three mutually perpendicular axes. These accelerations are then integrated twice, the first integration giving velocity and the second giving position. The system then directs the control system to preserve the pre-programmed trajectory.
- Astro guidance: this system constantly measures star angles and compares them with the pre-programmed angles expected on the missile's intended trajectory. The guidance system directs the control system whenever an alteration to trajectory is required.

11.3.1 Phases of Missile Guidance

In many missiles, the guidance system is divided into three phases, as shown in Figure 11.4. The first is a launch or boost phase in which the guidance system is usually disabled to allow the missile to safely travel away from the launch platform. The majority of the flights are flown using midcourse guidance, during which the missile makes slight adjustments to its trajectory allowing it to reach the vicinity of the target. The final phase is terminal guidance when the missile uses a highly accurate tracking system to make rapid maneuvers for

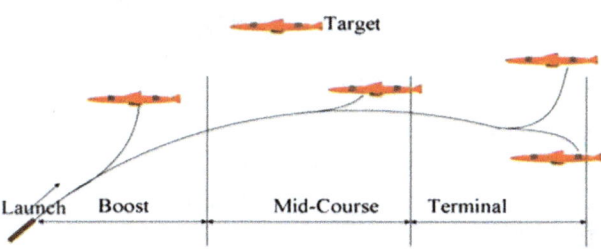

Figure 11.4 Missile guidance phases.

intercepting the target. Many missiles use a different type of guidance in the midcourse phase than in the terminal phase, and will be discussed later.

11.3.2   Missile Guidance Methods

The primary forms of missile guidance are described below with examples of missiles and seekers used to accomplish that type of guidance.

11.3.2.1   *Beam Rider Guidance*

The beam rider concept of missile guidance uses two ground stations: one for launching the missile and the other for locating radar for tracking and guidance. Radar tracks the targets and transmits the guidance beam, which adjusts itself according to the target movement. The missile contains a scanning system to detect the distance of the missile from the edge of the beam. The missile is launched in the guidance beams and it attacks the target riding on the guidance beam. The scheme is illustrated in Figure 11.5. Beam riding was often used in early surface-to-air missiles but was found to become inaccurate at long ranges and is no longer in use.

11.3.2.2   *Command Guidance*

Command guidance is similar to beam riding except for additional radar to track the missile itself. The tracking data from both radars are fed into a ground based computer that calculates the paths of the two vehicles. This computer also determines what commands need to be sent to the missile control surfaces to steer the missile on an intercept course with the target. These commands are transmitted to a receiver

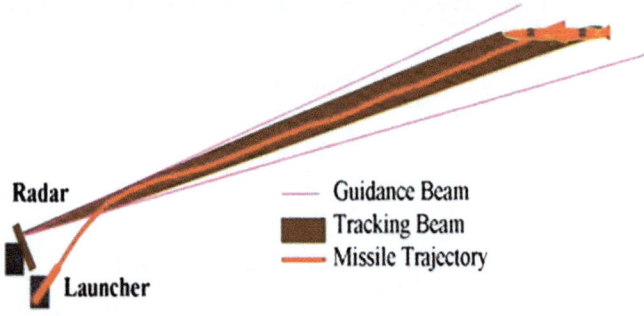

Figure 11.5   Beam rider guidance schematic.

on the missile allowing the missile to adjust its course. An example of command guidance is when the Russian SA-2 surface-to-air missile was used against a US aircraft in North Vietnam. Some of the wire guided missiles also use this guidance mechanism like anti-tank missile TOW, Naval Torpedoes *etc.*

11.3.2.3 Homing Guidance

Homing guidance is the most common form of guidance used on anti-air-missiles today. Three primary forms of guidance fall under the homing guidance umbrella: semi active, active, and passive (Figure 11.6).

In the semi-active homing guidance system, the target is illuminated by external radar signals and reflected signals are captured by the receiver of the missile. It is similar to command guidance with an onboard computer. The computer uses the energy collected by its radar receiver to determine the target's relative trajectory and sends correction commands to the control surfaces so that the missile will intercept the target. The air-to-air missile Arrow uses such guidance. Seekers used in this case are called bistatic because transmitted and reflected waves are different angles as compared to the line-of-sight between missile and target.

In active homing systems, the tracking energy is transmitted as well as received by the missile itself. No external radar source is needed in this case. It is for this reason that active homing missiles are often called 'fire-and-forget' because the launch aircraft does not need to continue illuminating the target after the missile is launched. Active homing systems are called monostatic because of the same inclination of transmitted and reflected signals from line-of-sight. Examples

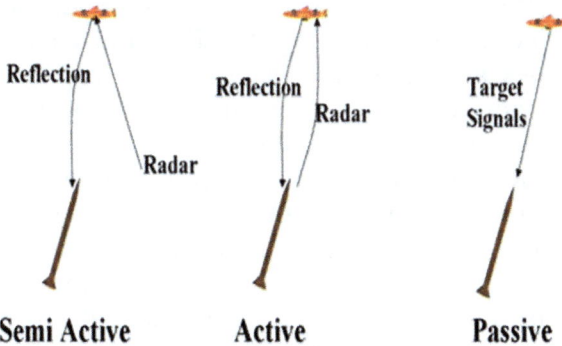

Semi Active   **Active**   **Passive**

Figure 11.6 Homing guidance variants.

of active homing missiles include the AMRAAM air-to-air and Exocet anti-ship missiles.

In passive homing systems, the missile receives signals from the target. It is independent of any external guidance system. Passive missiles instead rely on some form of energy that is transmitted by the target and can be tracked by the missile seeker. This energy could take many forms like infrared seeker homing to heat signature (in sidewinder), sonar dependent on sound waves (as in torpedoes), radio frequency energy (as in anti-radiation missiles), electro-optical signatures *etc.*

Another complicated method is the retransmission method. In this case, external radar is used for tracking the target, while reflected waves are received by the missile. However, the missile does not contain any computer and the reflected signal is again transmitted back to the launch platform for processing. Subsequent commands are then transmitted back to the missile for actuation of the control surfaces to adjust the trajectory. This method is also called track *via* missile (TVM). The advantage of TVM homing is that all of the expensive seeker and computer hardware is located on the ground where it can be reused for future missile launches rather than destroyed. Unfortunately, the method also requires excellent high-speed communication links between the missile and control station, limiting the system to rather short ranges. Retransmission homing guidance is used on the Patriot surface-to-air missile.

11.3.2.4 Navigation Guidance

Like homing guidance, navigation guidance includes several subcategories. In this section, inertial, ranging, celestial, and geophysical navigation techniques will be discussed.

Inertial navigation relies on devices onboard the missile that senses its motion and acceleration in different directions. These devices are called gyroscopes and accelerometers. While gyroscopes measure angular motion, accelerometers measure linear motion. When a gyroscope and an accelerometer are combined into a single device along with a control mechanism, it is called an inertial measurement unit (IMU) or inertial navigation system (INS). Inertial navigation works by telling the missile, where it is at the time of launch and how it should move in terms of both distance and rotation over the course of its flight. The missile's computer uses signals from the INS to measure these motions and ensure that the missile travels along its programmed path. Inertial navigation systems are widely used on all kinds of aerospace vehicles, including weapons, military aircraft, commercial airliners, and spacecraft. Many missiles use inertial methods for midcourse guidance, including AMRAAM, Storm Shadow, Meteor, and Tomahawk.

Ranging navigation depends on external signals for guidance. In earlier days it used radio signals. Based on the direction and strength of the signals, the location relative to the beacons could be calculated and navigation commands executed through the signals. The advent of the global positioning system (GPS) has largely replaced radio beacons in both military and civilian uses. GPS consists of a constellation of 24 satellites in geosynchronous orbit around the Earth. If a GPS receiver on the surface of the Earth can receive signals from at least four of these satellites, it can calculate an exact three-dimensional position with great accuracy. Missiles like JSOW and the JDAM series of guided bombs make use of GPS signals to determine where they are with respect to the locations of their targets.

Celestial navigation is one of the earliest forms of navigation devised by humans. Celestial navigation uses the positions of the stars to determine location, especially latitude on the surface of the Earth. This form of navigation requires good visibility of the stars, so it is only useful at night or at very high altitude. As a result, celestial navigation is seldom applied to missiles, though it has been used on many ballistic missiles like Poseidon. The missile compares the positions of the stars to an image stored in memory to determine its flight path. An ancient version of celestial navigation is geophysical navigation, which relied on the measurement of earth for navigation information.

A related but more accurate technique is called digital scene matching. Conceptually, digital scene matching is no different than looking out the window of a car and using landmarks to navigate your way to a specific location. Missiles make use of this technique by comparing the image seen below the weapon to satellite or aerial photos stored in the missile's computer. If the scenes do not match, the computer sends commands to the control surfaces to adjust the missile's course until the images agree. Digital scene matching is used in the Tomahwak cruise missile.

11.4   Laser Guided Missiles

Guided missile systems have evolved at a tremendous rate over the past four decades, and recent breakthroughs in technology ensure that smart warheads will have an increasing role in maintaining military superiority. On ethical grounds, one prays that each warhead deployed during a sortie will strike only its intended target, and that innocent civilians will not be harmed by a misfire. From a tactical standpoint, the military desires weaponry that is reliable and effective, inflicting maximum damage on valid military targets and ensuring capacity for lightning-fast strikes with pinpoint accuracy. Guided missile systems help fulfill all of these demands. Laser-guided missiles were first developed during the Vietnam War. The US Army began to research laser guidance systems in 1962. The first laser-guided bomb (BOLT-117) was developed by the US Air Force in 1967. However, it was not used in combat until 1968. The BOLT-117 worked using two planes. One plane was used to keep a laser illuminating the intended target, while the other's job was to drop the missile by following the reflected laser bean and directing the missile by sending signals to its control fins. For high efficiency, a very narrow region within which the pilot could release the missile was provided. Laser-guided missiles of this time were generally made of standard iron and were simply dumb bombs with a laser guidance and control system attached. They had a range of 3–4 km. Modern laser-guided missiles can be self-detonated, thus requiring only a single aircraft, and their range has increased significantly. The laser-guided missiles use lasers of a specific frequency bandwidth to locate the target. The pilot must line up the crosshairs and lock successfully onto target. This laser creates a heat signature on the target. The weapon must be released during a certain window of opportunity. After launch, the missile uses its onboard instrumentation to find the heat signature. The target is acquired, when the missile locates the heat signature. The missile is able to secure the target even if the target is moving.

Laser-guided missiles work by following the reflected light of a laser beam, which can either be shone on the target by the aircraft itself or by another airplane, or by ground troops with a hand held laser designator. Therefore, once the missile has been launched its own instrumentation is able to remain on target, rather than older laser-guided missiles that required the pilot to continually sight the target with the laser.

Laser-guided missiles are used for those targets that need pinpoint accuracy. A disadvantage of laser-guided missiles is that their guidance systems do not work well in all weather conditions. If the weather is cloudy, the water droplets in the air cause the laser to diffract. Because the laser only operates within a certain bandwidth, the laser can be completely diffracted, if it is too cloudy and the missile will not be able to locate its target. Rain has a similar effect on the laser because each raindrop serves to diffract the laser beam, once again deterring the missile from its target.

The precise work required by pilots sparked the development of other forms of guided missiles that do not require the pilot's guidance. Additionally, the weather limitations mentioned previously spawned a new breed of missiles that allow for accurate deployment in adverse weather conditions. Such missiles are guided using GPS technology. To guide such missiles, three coordinates are necessary, the latitude, longitude, and elevation. Developed by NASA in 2000, C-band and X-band interferometric synthetic aperture radars (ISFARs) are used to collect the topographic data required to employ this technology. NASA used these ISFARs to create the high-resolution topography of the Earth available today within ten days, with guided weapons being its primary application. These missiles have a longer range than typical laser-guided missiles.

Joint Direct Attack Munition (JDAM) missiles are based on a relatively new technology. JDAM is attached to the tail of the missile to change it from a conventional weapon to a GPS guided smart bomb. Accurate guidance is accomplished through a new tail section that contains a GPS aided INS, which guides the bomb from the release point to the intended target by the use of three coordinates. The INS with GPS updates allows the control fins to correct the trajectory until the moment of impact. These coordinates are sent to the bomb by way of an interface from the delivery aircraft. The missiles can be released up to fifteen miles from the target. JDAM also works in the adverse weather conditions that create difficulties in firing laser-guided missiles.

As discussed earlier, a missile may use one particular form of guidance throughout its flight or it may depend on different types of guidance at different times. Many weapons also make use of a combination of methods simultaneously. In particular, a common technique is a combined GPS/INS system that takes advantage of both inertial and ranging guidance to improve accuracy.

12

Special Topics in Rocketry

12.1 Thrust Vector Control (TVC)

The ultimate use of rockets is to deliver a payload to pre-defined targets. There are several uncertainties in attaining this objective like aerodynamic drag, gravitational pull, variable thrust, wind director, weather conditions and so on. For use of a system as a weapon it is mandatory to have high single shot kill probability (SSKP). Generally, three reasons are assigned for reduced SSKP: (1) improper take-off, (2) deviation from pre-defined trajectory and (3) movement of the target. To take care of these, a proper guidance system is devised for all the rockets, which calculates the miss distances and gives commands for steering the rockets towards their targets. As far as reference axes are concerned, angular rotation can be controlled along all three axes. In flight, any unwanted rotation about the longitudinal axis of a missile or along its flight axis is called rolling. If the nose of the rocket moves left and right about its center of gravity, it is called yawing and if the nose dives up and down, it is called pitching. Control on roll, yaw and pitch is called attitude control and it requires the generation of controlling the torque for stabilization of flight. During the propulsion phase, controlling torque is usually developed either by aerodynamic forces acting on the control surfaces or by rocket forces. It needs a reference, from where error is to be measured, a detector which measures the error and a controller, which takes corrective action to minimize the error from the reference.

Solid Rocket Propellants: Science and Technology Challenges
By Haridwar Singh and Himanshu Shekhar
© Haridwar Singh and Himanshu Shekhar 2017
Published by the Royal Society of Chemistry, www.rsc.org

Among the various methods of control, aerodynamic Cartesian control utilizes a moving wing, which diverts the push on rockets towards a reduction of miss distance. Similarly, aerodynamic polar control and autopilots have been utilized in various missiles. However, an emerging trend is to utilize Thrust Vector Control (TVC) for the maneuverability of rockets and missiles in the correct direction.

Aerodynamic controls have been utilized in conventional aircrafts but rarefied air (low density) at high altitudes makes this system ineffective and other means of producing controlling torque is advocated. To reduce load on airframes of large size motors, gentle vertical launch is preferred. In this case, aerodynamic control is not needed as it may topple the missile, if used in the early stages of flight. In the case of TVC, the thrust axis is turned with respect to the direction of motion of the rockets. The angular deviation causes the generation of side thrust at the cost of axial thrust, resulting in a moment for giving pitching or yawing. This control has been effective for large sized rockets to maintain their correct trajectory. Another problem with aerodynamic control is the achievement of full speed before attainment of full maneuverability. For short range air-to-air missiles, it is difficult to wait until full speed is attained for engaging a target. In this case TVC is also preferred. There are several methods to achieve TVC in rocket motors.

In case of rocket motors based on liquid propellants, the combustion chamber can be mounted on gimbals and rotated by servo to achieve TVC. This device can be operated at any missile speed and is independent of the external atmospheric conditions. Such a gimbaled combustion chamber needs feeding of its propellants through flexible pipes without the creation of any shock waves or loss of efflux momentum. Swiveling of the entire rocket motor was planned for the flight of the Thor rocket. However, for solid propellants, a ball and socket nozzle or flex nozzle with deflection angle a little above ±15° is preferred. However, this makes the nozzle heavy and friction at the seals requires very powerful servos. A ball and socket nozzle was incorporated in the Minuteman missile. Shuttles have used flexible bearing nozzles during their flights.

In liquid propellant rockets, thrust can be controlled by the use of gimbaled thrust chamber and the direction of thrust can be changed at will. Cutting off the valve controlling the propellant supply can also terminate thrust. For solid propellants, the use of jet vanes and jetavators are well suited. This basic difference between the two systems lies in the fact that the jetavator is exposed to hot exhaust gases only during the controlling operation. However, both systems increase the

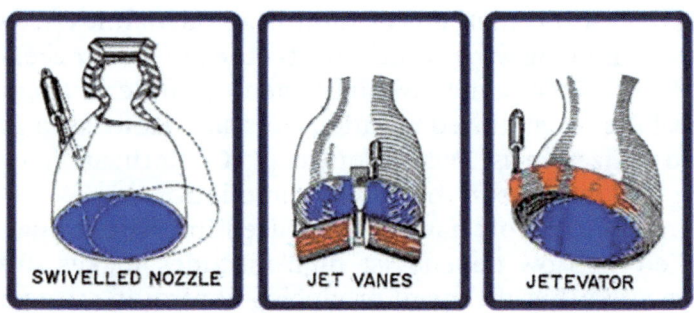

| SWIVELLED NOZZLE | JET VANES | JETEVATOR |

Figure 12.1   Methods of controlling thrust in solid propellant rockets.

weight of the nozzle assembly, but jet vanes are more effective. They are lighter in weight than jetavators. A third method of control is a movable exhaust nozzle. If one can ensure a gas seal, the method is very promising. Figure 12.1 shows three possible combinations.

Another scheme would be the opening of a forward facing port to equalize the forward thrust. With the development of a satisfactory thrust control mechanism and a positive means of terminating thrust, solid propelled rockets have become more competitive than liquid propellant rockets.

In typical situations, spoiler systems have also been incorporated, where spoilers deflect the outgoing jet from the motor nozzle with the creation of shock waves. This results in a reduction of thrust by about 1% per degree of thrust deflection. However, this needs smaller servos due to the smaller size of the moving parts. Jet vanes (used in Jupiter, Redstone, V-2 & Juno flights) and Jetavators (used in Polaris and Bomarc flights) operate with low auxiliary actuator power and provide a high turn rate. However, there is a thrust and Isp loss up to 2%. Jet vanes are placed in the outgoing stream of hot gases from the motor and their life is limited by erosion coupled with exposure of high temperature. The life of the jet vanes in the German V-2 rocket was around 60 s. In Redstone and V-2 missiles, both carbon-jetvanes and aerodynamic control surface were assembled. Initially, when the missile speed is zero, jet vanes provide stabilization and later aerodynamic control picks up. Jetavator is in the form of a ring around the circumference of the motor nozzle. In the neutral position, it does not cause propulsion losses and on requirement this ring deflects the exhaust. It is frequently used for solid rocket motor attitude control.

Another popular mechanism is Secondary Injection Thrust Vector Control (SITVC), in which a secondary fluid is injected in the nozzle divergent to create oblique shock. This provides efficient control and increased motor performance. However, additional liquid propellant and injection equipment are needed for such operations. In an actual scenario, pyroseal valves seal injectants prior to ignition. Exhaust gases melt the exposed end of the valve allowing the injectant to flow in the outgoing stream of hot gases through the nozzle. In the Titan family of rocket motors, nitrogen tetroxide (N_2O_4) was used as an injectant.

Yet another approach is incorporated in satellites for altitude correction or roll correction. Small auxiliary thrusters are used for this purpose. This approach is explored and practically demonstrated in Magellan flights. The mono-propellant attitude control system is designed to provide three axis control with or without the solid motor firing. It provides both pitch and yaw control during firing. In the Pioneer lunar probe, the vehicle is spun at propulsion cutoff in an orientation that enables it to be placed in orbit around the Moon by a reverse thrust rocket.

Earlier rockets employed 'jet vanes', which are graphite vanes located 90° apart in the gas stream of a single nozzle and usually coupled mechanically to aerodynamic fins external to the nozzle. In the environment of high temperature, pressure and flow velocity due to increases in the propellant burning temperature and upgrading of propellant performance, graphite vanes erode, burn and crack. Hence, new methods of TVC were sought.

The TVC systems developed operate by deflecting the flow direction of propulsion gases by deflecting gases after expansion, by deflecting the nozzle or by deflecting the gas stream within the nozzle. Examples of these three types are called Jetavator, movable and secondary injection nozzles, respectively. A few methods of jet control are illustrated in Figure 12.2. Both shock-free and shock-augmented systems have been developed and are currently being used in missile systems. Mission and deployment of missile influence the choice of a specific TVC.

12.1.1 Shock-Free Thrust Deflection System

The shock-free system operates on the principle of redirecting the gas flow in a low Mach number regime ($M \leq 0.5$). Turning at this low Mach number eliminates shock losses, simplifies heat transfer considerations, and affects the flow coefficient of the nozzle slightly. Some inherent disadvantages of these TVC systems are the complexity and vulnerability of the primary seals, high inertial loads that the

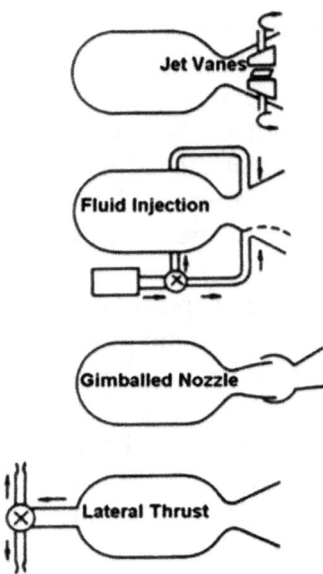

Figure 12.2   Some methods of jet control.

actuation system must overcome, and the unpredictable friction loads due to thermal expansion and pressure deflections. However, all of these disadvantages can be overcome by appropriate development. Because of the low performance losses of these systems, some weight penalties can be accepted for the inherent reliability and accurate controllability of the shock-free movable nozzles. These shock-free TVC nozzles offer the best over-all nozzle efficiency for any thrust vector control system, but they are not the lowest thrust loss system, as will be discussed in the section on fluid injection systems.

12.1.2   Shock-Augmented Thrust Deflection Systems

Shock-augmented TVC devices can be subdivided into mechanical systems and fluid injection systems. Supersonic flow shock phenomena are used in both types of systems to generate an additional imbalance of pressure, thus augmenting the side force. The losses sustained by changes in entropy due to shocks vary among the different systems, because the intensity of the shock varies. In some of these systems, the shock strength is low enough to allow the isentropic compressive-turning equations to be used as a first approximation during design calculations. Others, however, require detached strong shocks in order to function effectively.

12.1.2.1 Mechanical Systems

The mechanical system in which shock augmentation is used for thrust vector control include movable expansion cones in the nozzles, jet tabs, jetavators, jet vanes and spike injection. In all of these systems, the mass flow of the propellant gases is constant because all TVC occurs in or aft of the expansion cone and therefore cannot influence flow rate of the motor.

a. Movable expansion cone nozzle: in the nozzle with movable expansion comes either a pivoting or offset-rotatable expansion cone that is used to divert the flow leaving the nozzle exit plane. The advantages of this system are the rapid response capability due to the low inertia of the exit cone; the increase in reliability because of low pressure moving seals can be used, as opposed to high pressure seal in the shock-free movable nozzles; the improvements in reliability and packageability because a stationary nozzle throat is used and can be buried in the motor; the ease of actuating the system because ejection to thrust loads can be balanced on the movable portion; and the relatively light weight. The disadvantages of this system are the energy losses due to shock flow condition; the limitation on vector deflection angle (10–15° maximum) before detached shocks occur; the unsymmetrical pressure loading and thermal inputs to the expansion cone, which require refractory insert materials in this area and nullifying ablatives; and the loss of side force-producing capability with changes in contour due to erosion and deformation.

b. Jetavators: these are contoured, ring shaped devices that pivot on two bearings mounted near the exit plane of the nozzle. This device was developed and used successfully on early Polaris motors. The term 'jetavator' is a contraction of 'jetelevator' and describes the function of the device; elevating and redirecting the jet streams. Forces to actuate the jetavator are composed of frictional, inertial and aerodynamics loads, the latter normally being the largest. The jetavator was a highly reliable TVC system for early solid propellant rocket motors.

c. Jet tab: the jet tab is a contoured, plate-type structure used to locally block a portion of the nozzle exit plane, thus producing a strong shock and side force in a manner similar to the jetavator's. Up to eight of these tabs can be used on one nozzle and, if properly controlled, can produce roll control as well as pitch and yaw control. Once again, erosion is undesirable because changes in contour affect vector control ability and actuation forces. The device is being seriously considered for very large solid fuel motors because of its simplicity, TVC capability, reliability and relative light weight. The low inertia and relatively low friction and aerodynamic loads on this device allow the use of simple, lightweight actuation systems. This device is desirable for use on large boosters, when the relatively large impulse losses due to shocks are not very important.

d. Jet vanes: the jet vane is an aerodynamically contoured vane, normally located in the exhaust stream and mounted on a pivot at the nozzle exit plane. The major advantage of jet vanes is roll control capability and rapid response with no 'dead-band' between zero actuation and full deflection. This is a high loss system because some shock losses are present even when the vanes are not deflected. However, the system is lightweight and relatively simple. It was one of the first TVC systems used in early rocket motors such as the German V-2 missile. Jetava-tors and jet vanes have had a long, successful period of application but are currently considered not applicable to modern high performance motors. The jet tab, as previously stated, has some potential for large booster work because it incorpo-rates the best features of both vanes and jetavators, without the more serious drawbacks. All of these are high-loss systems requiring a strong shock for effective operation and are there-fore considered only for current applications, where speed of development is more important than highly sophisticated performance and design.

e. Spike injection: the spike injection system consists of a shaped refractory spike that is inserted into the stream through the wall of the exit cone and produces a shock, whose strength varies with the amount of spike immersed. Very accurate and rapid response TVC is possible with this system by appropriate shap-ing of spike and thus control of the shock. Because the spike is a relatively slender unit, the standard heat-sink principle of surface-temperature propellant gas requires cooling of the spike for any sustained immersions. It is possible, however, to use the cooling medium as the actuation fluid, thus saving considerable weight. This system is inherently lighter than jet vane, jetava-tor, or jet tab as a pitch and yaw force producer, because of the small additional structure required for mounting. However, this system does not provide roll control. The complexity of the cool-ing-actuation system and the availability of numerous other TVC systems has somewhat retarded the development of this system.

Spike injection has a good potential for application in missile systems, where only pitch and yaw control, very high acceleration, and very high TVC response rates are required. The losses in this system are shock losses due to entropy change, and the maximum side-force capability is limited to effective angular deflection of about 10°.

12.1.2.2 Fluid Injection Systems

Fluid injection systems operate on the combined principle of side jet flow and shock augmentation. Because fluid is injected at a large angle to the main nozzle flow, some side thrust results because of mass ejection effects. This side thrust is augmented to a large degree by the shock formed when the secondary stream impinges on the main stream. Both reactive and inert liquids and gases have been used as secondary fluid injectants. In their order of increasing effectiveness (side thrust/mass flow of secondary injectant), the secondary fluids are (1) non-reactive liquids (2) reactive liquids (3) non-reactive cold gases (4) reactive cold gases (5) non-reactive hot gases and (6) reactive hot gases. In fuel-rich exhausts, the reactive fluids would be oxidants. Conversely, in oxidizer-rich exhausts, the reactive fluids would be fuels. Monopropellants, such as 90% hydrogen peroxide, have been used with good effect. The choice of the fluid depends on the volume, weight, storage environment, and safety regulations of the missile system. The best fluid to produce side specific impulse often cannot be used. The basic advantage of any injectant system as a TVC device is that the injectant fluid augments the main thrust. In terms of total missile performance, this system carries its own weight to some extent and is therefore less subjected to the weight penalties common to other kind of systems. This feature is important for upper stage motors, where inert weight causes severe payload or range penalties. In booster stages, where large side forces are required, any mission trade-off studies must be run to choose the best TVC system, with fluid injection suffering the handicaps of complexity, lower inherent reliability, and loss of efficiency at high flow rates. Where deflection requirements exceed four or five degrees, fluid injection does not compete well with the mechanical systems; but this may change when reliable proportionate hot gas valves are developed.

12.1.2.3 External Burning

The possible application of the external burning principle falls into three classes (a) side-force generating devices for attitude control,

(b) thrust generating (or drag reducing) devices, and (c) devices that produce both thrust and attitude control or lift. Initial experiments in external combustion using hydrogen as a fuel encountered difficulties in establishing ignition and combustion. For this reason, most experiments have used very reactive fuels including aluminium borohydrides, triethyl aluminium (TEA), pentaborane, and triisobutyl aluminium (TIBA). Experimental results indicate that performance may be reaction-rate limited. Aluminium borohydride appears to be the fastest reacting fuel tested, followed by pentaborane and the aluminium alkyls TEA and TIBA. Higher pressure generally increases reaction rates; therefore lower altitude should require less strength. Compression prior to injection due to the body shape and/or burning on the windward side at angle-of-attack will help because the local pressure and temperature will be higher and the velocity somewhat lower. Lower flight velocity increases the residence time for a given length but also has associated with it lower maximum temperatures in the boundary layer which may be important. Finally, the wall temperature and the fuel temperature, which could be elevated if the fuel is used as a regenerative coolant, may also be important.

12.2 Structural Integrity of the Propellant

Structural integrity (SI) still remains an area of concern to ensure performance reliability and fulfillment of mission objectives. At every phase of its life, the solid propellant grain is subjected to environmental and operational forces, which try to degrade its structural integrity. In almost all cases, the requirement to withstand longitudinal acceleration in flight of 40–100 g is normal. In future, motors will be required to accelerate greater than 100 g. In recent years, capability of aircrafts to cover a wide range of altitudes at supersonic speed has increased the stresses imposed on propellant grains due to thermal cycling and aerodynamic effects. The cycle of take-off, flight and landing is repeated in peace time. The structural integrity of propellant grains therefore must be adequate to resist these stresses and produce reliability of 0.999 for safe performance.

During flight, high Mach numbers generate stagnation heating conditions, which add further to those imposed by flight conditions. For long burning case bonded motors, interface effects caused by high temperatures between liner, case and propellant can be very serious and may cause separation or debonding, leading to catastrophic failures. A lot of improvements have been made in recent years in the formulations and technology of insulators, liners and case materials.

Furthermore, the strength of propellant has been improved considerably due to the usage of powerful bonding agents. Spin acceleration is another factor, which demonstrates unusual effects on the performance of rocket motors. High spin rate affect burning rates and gas dynamics of the propellant. This may result in serious erosion effects and aberrations in thrust delivered.

There is always a need for complete characterization of time and temperature dependent mechanical properties of the propellants as a basis for determining structural integrity. Each propellant system has its own characteristics. Double base propellants are very useful in a free standing mode, but conventional DBP propellants can not be case bonded. Composite propellants based on PBAN/CTPB/HTPB offer promising physical properties over a wide temperature range. However, optimization of mechanical properties of propellant becomes very important for a particular mission. In this area, polymer (binder), molecular weight, molecular weight distribution, unsaturation, chemical structure (*cis*, vinyl, *trans*) and cross-linking plays a very important role. Various test methods used for mechanical properties characterization and optimization include: uniaxial test method, constant strain test, stress relaxation modulus test, multi-axial test method, thermal stress evaluation, bond test *etc.* Pull off test and double lap shear tests are used to evaluate resistance of propellant-liner-metal bonds to failure, resulting from shear forces.

While undergoing thermal expansion and contraction during temperature cycling, the propellant grain must not be brittle at low temperature or else it will crack, neither must it be too soft at high temperatures to deform. In case bonded motors, the differential expansion of motor case and propellant adds to the stresses at the interface. If the geometry of the propellant is complex, it leads to a point of stress concentration. For certain applications, even case bonded propellants with TS more than 4 kg cm^{-2} and elongation more than 20% have been quite useful.

The physical properties of double base propellants (DBP) are a reflection of the viscoelastic behavior of NC, modified by plasticizers, both energetic and conventional. If the polymer is linear, partly crystalline, highly polar and cross-linked, the viscoelastic behavior becomes complex due to both elastic and plastic deformations. Elastic deformation has a time dependent impact. Hence, a maximum TS, elongation at maximum stress and modulus of elasticity must be measured at extreme temperature before a propellant is selected for application. In DBP, the type and content of NC plays very vital role. The effect of NC percentage on TS and elongation are shown in Figures 12.3 and 12.4. The data is for NC of 12.6% nitrogen. Low

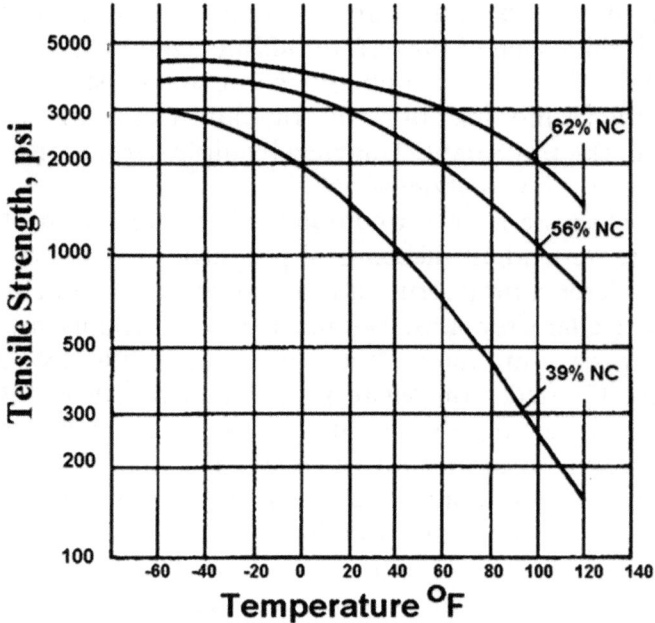

Figure 12.3 Effect of %NC on tensile strength in CDB propellants.

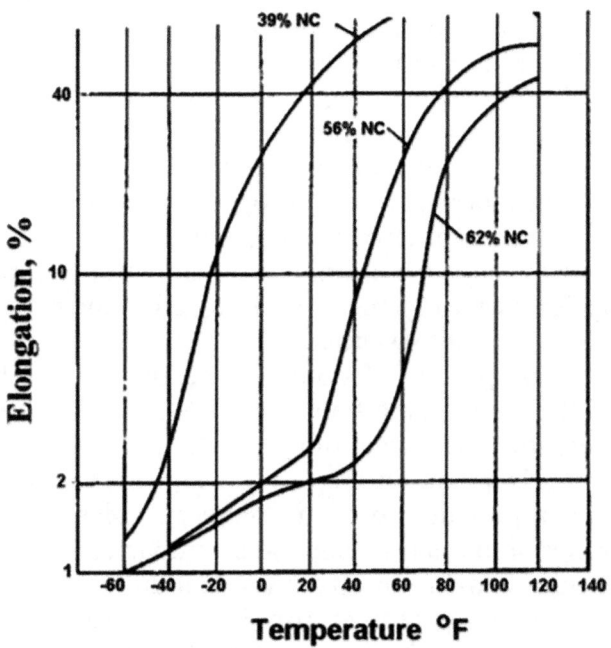

Figure 12.4 Effect of %NC on elongation in CDB propellants.

nitrogen content NC (12.2% NC) produces higher mechanical properties than NC Pyro (12.6% NC) and GC (Gun cotton) having 13.1% nitrogen.

In case of composite propellants, considerable work has been done to relate propellant properties with the viscosity of the fuel cum binder, volumetric concentration of oxidizer and particle size distribution of the suspended oxidizer. In addition to the concentration of suspended particles, the particle size distribution is also very important.

An important component of composite propellant thermo-visco-elastic behavior results from the interaction of the rigid oxidizer and the fuel particles with the soft binder material. This interaction manifests stress or deformation as dewetting or debonding from the particle surface and binder at peel stress locations. These microscopic stress risers are the origin of propellant mechanical failure. Figure 12.5 shows the results from a low temperature uniaxial strain dilatation test on a composite propellant specimen. As shown, dilatation at large stress levels near failure can easily reach 15–20% volume increase during a uniaxial test. This level of material void increase during loading is a clear indication of a departure from continuum material behavior even at moderate strain levels. The consequences of this microstructural behavior combined with the binder's thermo-viscoelastic effects are significant and difficult in modeling both constitutive and failure behaviors for structural and margin of safety analysis.

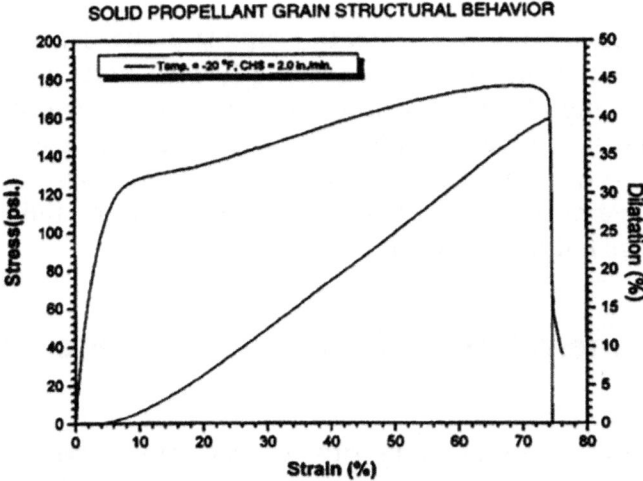

Figure 12.5 Dilatation behavior for composite propellants.

A grain structural integrity analysis is an evaluation of the capability of the solid rocket motor to meet the mission requirements under the specified environment. The grain structure analysis determines the stresses, strains and deformations that the solid propellant grain will experience under the specified environment. The structural analysis is then coupled with a failure analysis to define the environment in which the motor will operate successfully. Since operational conditions are generally well defined in the beginning of the program, analysis is used to determine a minimum factor of safety of motor operation. The data required to conduct a grain structural integrity analysis are the definition of loads that the fielded or stored motor will experience, component material properties, failure criteria and required factor of safety.

Transportation and handling loads encompass loads imposed by vibration during transportation, vibration under the wing of airplane for air-launched missiles and shock loads imposed by dropping missiles during transport. The vibration loads, which are often of a debonding mode, generally induce very small stresses compared to thermal, acceleration and pressure loads but these should be considered in combined loads failure analysis.

Pressurization loads occur during ignition and continue throughout burning of the propellant grain during flight. These loads impose hydrostatic compressive stresses on the propellant grain ends. When ignition starts, the pressure that is typically built rapidly to a maximum level, induce critical load. The propellant properties, grain design and motor case stiffness determine the imposed pressurization stresses and strains.

When applying failure criteria to propellant charge, consideration must be given to state of stress, loading history, temperature history, chemical ageing effects and finite deformation. It is very important to note that no universal failure criteria have been established for propellant grain. The material property changes that occur during chemical ageing may result in a change in failure mode, thereby changing the failure criterion that is appropriate.

12.3   Modern Rocket Motor Casing Materials

Rocket motors are generally made of metallic materials or fiber reinforced plastics (FRP). The majority of rocket motor casings are made off either titanium alloy, or a composite filament wound structure such as Kevlar-epoxy or S-glass-epoxy. Titanium alloys need a long lead time for procurement of forgings. The fabrication time is also longer due to complex machining, welding, heat treatment and final precision machining. Fabrication by strip welding of a typical metallic motor case of 1 m diameter and 5 m length also takes longer. High strength composite motor cases need thicker walls but overall weight remains less than the metallic casing weight for the same size and pressure capabilities. Much lighter and more efficient motor casings are being made by filament wound graphite epoxy (fiber strength 800 ksi).

Critical issues in the selection of rocket motor casing materials have been the availability of material as well as that of established fabrication methodology. In addition, uncertainty in the joint region of composite casing coupled with a lack of cost advantage also plays a major role. The motor case hardware material must have following pre-requisites:

- High material yield and ultimate strength.
- High weld efficiency (min. 90%).
- High fracture toughness (>90 MPa $\sqrt{m}$).
- Ease of machining, welding, forming.
- Expertise and fabrication infrastructure developed by the industrial base.
- Simple heat treatment cycle.

Among the conventional structural materials used for rocket motor cases are good quality alloys of steel, aluminium, and occasionally titanium. These materials are homogenous with isotropic properties. Table 12.1 compiles the properties of promising materials along with the relative mass factor calculated on the basis of both stress and strain criteria.

On a stress basis, the standard Al-alloy is 7% heavier than high tensile steel but all other materials show a reduction in weight. The maximum reduction in weight is observed in case of maraging steel (35% reduction) among metals. On a strain basis, all metals are more or less similar. However, the highest weight advantage is obtained in the case of Al-alloys (8% reduction). Rocket motors are generally designed on the basis of stress criteria and maraging steel seems to be the obvious choice for the same. Contrary to this, end closures and blast tubes are designed on the basis of strain and Al-alloys becomes an obvious choice. It is not only a reduction in weight, other criteria like aerodynamic heating, economical manufacturing, thickness to diameter ratio also affect materials selection for various propulsion units. Table 12.2 summarizes the properties of rocket casing materials used in various systems globally.

Table 12.1 Properties of rocket motor casing materials.

Material	Density (ρ)	TS (σ)	Modulus (E)	Mass factor for pressure loading relative to high tensile steel	
				Stress 17.83 $\times \rho/\sigma$	Strain 2547.77 $\times \rho/E$
	(g cm^{-3})	(kgf mm^{-2})	(kgf mm^{-2})		
High tensile steel	7.85	140	20 000	1.00	1.00
Heat treated steel strip	7.85	200	19 300	0.70	1.04
Maraging steel	8.00	220	21 500	0.65	0.95
Standard Al-alloy	2.70	45	7300	1.07	0.94
Heat treated Al strip	2.70	54	7300	0.89	0.94
Latest Al-alloy	2.70	60	7500	0.80	0.92
Titanium forgings	4.65	124	11 000	0.67	1.08
Heat treated Ti strip	4.65	135	10 600	0.61	1.12
Glass fiber/Epoxy	2.10	63	2400	0.59	2.23
Aramid fiber/Epoxy	1.40	87	3600	0.29	0.99
Carbon fiber/Epoxy	1.60	100	7300	0.29	0.56

Table 12.2 Material of construction of some rocket casings.

Propulsion systems	PS–I	Titan	Minuteman	Space shuttle	Ariane	H–II
Country	India	USA	USA	USA	France	Japan
Material	Maraging steel–250	D6AC	D6AC	D6AC	48CDN 4–10	NT-150 steel
UTS (kgf mm^{-2})	175	150	150	150	150	150
Yield (kgf mm^{-2})	170	130	130	130	130	130
Fracture toughness (MPa $\sqrt{m}$)	90	—	—	100	85	—
Heat treatment	Ageing 480 °C	Harden and temper	Harden and temper	Harden and temper	Harden and temper	Harden and temper

End domes, igniter boss, holders and nozzle joints are designed based on material selection and available manufacturing technology. Since end domes are zones of high discontinuity stresses, design and fabrication has to match the requirements. Spherical domes being deep require more space, resulting in a weight penalty. They are superior from the point of induced stresses. Ellipsoidal domes are relatively flatter but pose problems in fabrication due to a varying meridional radius. Tori-spherical domes, consisting of separate knuckle and crown radii are preferred for ease of manufacturing. However, they buckle under internal pressurization and compressive hoop and tensile meridional stresses are induced in the knuckle region. In space shuttle programs, end-domes were machined out from large forgings but forming route is relatively simpler and is preferred manufacturing route.

For the fabrication of rocket motor casings of homogenous materials, several manufacturing tools are available. Welding is one of the major process for manufacturing tubular casings. For large sized motors, longitudinal welding gives a tubular casing of limited length, which is coupled with circumferential welding to enhance length. A combination of these is possible in spiral winding, where strips are welded along a helical path to obtain cylinders of any length and diameter. For components with a low ratio of length to diameter, forging is employed for production. End rings, end domes, and blast tubes can be readily made by this method, which gives good grain flow. Casting is an economical but not a very popular method for production of rocket motor casing. For components having low length to diameter ratios made of medium to high strength steels, deep drawing is preferable. For ductile materials with relatively lower strength, extrusion is employed. This is limited to low diameter aluminium components. Flow forming is fast gaining importance in rocket motor case production industries. Flow forming gives higher strength in hoop direction, which is advantageous for pressure vessels. It can be readily used for aluminium alloys and has been used for the fabrication of 200 mm diameter rocket motor hardware of SAE 4130 steel.

The improvement in material properties of aluminium is possible by alloying it with lithium, which results in a light weight material with higher modulus. A major problem with metallic components is weathering or deterioration in properties on exposure to the environment during service life. Since most of the systems use either steel or aluminium alloys as casing materials, external surfaces are generally painted and tested for protection against salt spray, acid rain, prolonged exposure to hot wet conditions, mechanical abrasion, thermal heating *etc.* Cadmium plating is used for enhancing the corrosion resistance of steel surfaces. This is done by electroplating in a very cost effective way. However, this process with high strength steel causes hydrogen embrittlement. Cadmium has limited protection against salt spray and is toxic in nature. A phosphate coating can also be done in certain situations. Stainless steel and titanium can be left unprotected and have shown no deterioration on prolonged exposure.

Another strength enhancement mechanism is strip laminates preparation. All the materials given in the table show enhanced strength in strip form. There is no limitation on strip materials, so long as they possess sufficient ductility for winding. To reduce winding time, wide and thick strips are preferred. However, due to lateral curvature and gap variation between successive turns of a helix, the width of the strips is limited. Generally, strip width is less than the tube diameter and the thickness of the strip is dictated by ease of winding. For bonding layers of strips, epoxide systems have been developed and used. The adhesive strength achieved is generally more than 3.5–4 kg mm^{-2} for practical systems.

The latest trend has been the replacement of metallic casings by composite materials at least for the upper stages of launch vehicles. The effort for this change is driven by the availability of the material, mechanical properties database, established fabrication methodology, proper interface and joint design and proven non-destructive testing expertise. The major advantages offered by composite casing is a high strength to weight ratio compared to its metallic counter part. The normally used composite materials for realizing large motor casings are fiber glass, Kevlar and graphite (carbon) with matrix materials like epoxies. The properties of typical materials are given Table 12.3.

Initially fiber thickness is ascertained and the number of hoop and helix layers are arrived at by selecting the proper layer thickness. The important aspect of composite casing design is a judicious use of directional properties of composites to realize a very efficient low

Table 12.3 Properties of composite case materials.

No.	Properties	KEVLAR/Epoxy	Carbon/Epoxy
1.	UTS (kg mm^{-2})	106	146
2.	Modulus (kg mm^{-2})		
	Along fiber	5530	13 020
	Transverse to fiber	300	601
3.	Failure strain (µε)	20 000	12 000

weight motor casing. Composite casings are generally realized by a filament winding process, which involves placing alternate layers of helix or polar wraps with 90° or hoop wraps onto a mandrel shaped to form the desired case interior profile. The polar wrap carries all the axial membrane loads, while hoop wraps bear the tangential membrane load in the cylindrical portion. For the domes, the polar opening diameter and the cylindrical section diameter dictates the angle the helix band must have, as it crosses the dome-cylinder junction. Once the winding is complete, the entire set up is cured in an autoclave. This is followed by a mandrel extraction operation, leaving behind a rigid light weight motor case.

Composite materials are obvious choices from the mass factor considerations presented in the Table 12.3. On a stress basis both C-fiber/epoxy as well as aramid fiber/epoxy have a high reduction in weight (71%) whereas on a strain basis, C-fiber is the obvious choice. For practical utility, other constraints must be addressed properly. In composites, a reduction in density coupled with a high strength results in a lower mass fraction requirement but casing thickness generally increases. For the C-fiber/epoxy system, the stress criteria gives a thickness multiplier compared to high tensile steel as 1.4 times on stress basis and 2.74 times on strain basis. In addition to this, dilatation of the composite casing during internal pressurization has to be controlled specially for case bonded application. A higher distortion of the casing may result in interface debonding.

12.4   Life Prediction and Extension of Propellants and Rocket Motors

A useful life of a propellant is one of the major criteria for selection and usage of propellants/rocket motors for a particular application. Useful life ensures the achievement of mission goals and safe handling during processing, transportation and storage. Propellant life is generally classified under three main categories: (1) safe storage life or shelf life, (2) safe use life and (3) useful life. As the name indicates, it is the useful life that assumes a high importance from an application/mission point-of-view. The useful life is the period beyond which the propellant or rocket motor cannot be expected to perform its mission reliably, although it may be safe to store in the magazine/depot. During aging, the propellant may undergo deterioration in chemical composition, mechanical properties or ballistic properties or a combination of all three.

The major effects of aging include migration of propellant ingredients towards the inhibitor/insulator or liner, loss of explosive ingredients content, deterioration of the bonds between propellant and inhibitor/liner and loosening of bonds between the propellant and the motor casing material. With regards to the mechanical properties, there may be a change in the Ultimate Tensile Strength (UTS), Compressive Strength (CS), Young's modulus, *etc.* leading to the generation of voids and cracks. Changes in ballistics include a change of burn rates at different pressures, loss of energy (Isp), unstable combustion including pressure oscillations and changes in the temperature sensitivity coefficient. Some other important factors that may affect the useful life of a propellant include storage conditions, humidity, temperature cycling effect, auto-decomposition of propellant ingredients, crack formation and their growth, material separation, diffusion of ingredients, stress–strain influences *etc.*

In the family of solid rocket propellants, double base propellants (DBP) consisting of nitro-cellulose (NC), nitro-glycerin (NG), stabilizers, plasticizers and other additives have a long storage life of 20–25 years. In view of longer useful and safe storage life, they have been extensively used for various military applications, where higher energy is not of prime concern. Composite propellants (CP) comprising of an oxidizer (AP/KP) and/or high energy materials like RDX/HMX and metallic fuel (Al) and polymeric binders (HTPB/CTPB), which act both as a fuel and a binder produce higher energy than DBP, (Isp > 240 s). They have comparatively lower life and lower mechanical properties. However, unlike DBP, composite propellants produce stable combustion at low pressures (15–25 kg cm^{-2}). A drawback of composite propellants is that aluminized propellants produce smoky exhaust products, which may affect the guidance system of missiles.

Composite modified double base propellants (CMDB) take advantage of both a fuel-rich double base matrix and an oxygen-rich composite propellant and therefore, are capable of producing higher energy (Isp > 260 s) and superior mechanical properties than composite propellants. In view of the high potential of CMDB propellants in terms of energy (Isp) and superior mechanical properties, they are being increasingly used for tactical and strategic missiles. The cross-linking of un-nitrated hydroxyl groups of NC in CMDB propellants further enhances the mechanical properties. A new class of CMDB propellant, where a nitric-ester soluble binder is included in the matrix, is emerging in a big way, which is very attractive for low temperature applications and can be case bonded. This class of propellants are generally known as elastomer-modified composite modified double base propellants (EMCMDB). An Isp of the order of 280 s and burning rates of the order of 50–200 mm s^{-1} have been claimed for CMDB propellants.

While a number of studies have been conducted on aging behavior of DBP and CP, limited studies have been carried out on the aging behavior of CMDB and advanced solid propellants containing newer energetic materials. Moreover, most of the studies are only restricted to bare propellant and hence may not be useful in predicting the total life period of practical motors, as most of the propellants (irrespective of their class) are either inhibited to obtain neutral thrust–time profile or case bonded, where the liner/insulator quality also plays a crucial role.

The life extension exercise of aged motors has assumed high economical importance these days, due to huge amount of financial saving, in view of the fact that most of the developers/suppliers of rocket motors generally assign very conservative life estimates (10–12 years).

The essential difference between the serviceability and life extension of rockets/missiles is that in the case of the former, only limited tests are carried out and the deterioration of the explosive component is assessed for a short time. In this process, it is presumed that rate of degradation or deterioration is not sudden but is uniform over a period of time, depending upon the storage/deployment conditions. However, in the case of latter, a large number of tests are required to be carried out on both life-expired and active missile propellants.

12.4.1 Failure Modes

Various failure modes are possible during ageing and include propellant grain cracking by thermally induced stresses, pressurization, vibration, shock, case-liner joint separation *i.e.* debonds, acceleration, improper processing, ignition failure *etc.* The test methods for determining the failure modes of a system include testing of full scale or half version motor and prediction of performance by finite element analysis. However, this approach works out to be costlier, in view of testing the motors at various temperature conditions. Some failure information may be available through a developmental database of a system due to the occasional failure of the motors observed during service and their correlation.

Accelerated aging is generally adopted to predict the life of a newly developed rocket motor. However, this approach suffers from the limitation that it does not reflect a real life situation, as rocket motors are generally stored in air-conditioned storage, having control of humidity and temperature. Many times the life predictions made following an accelerated aging cycle and determining the rate of deterioration is much lower than the actual life. During accelerated ageing trials, rocket motors are subjected to accelerated ageing at high temperatures (+50–60 °C) and ageing under normal conditions of storage (+20–25 °C) under humidity control.

In real time assessment, the stored rocket motors at elevated temperatures are tested for propellant structural integrity and ballistic behavior over a period of time (10–20 years). This scheme involves periodic sampling and generation of data and comparison with base line data (fresh rocket motors). Although this method is very attractive, but is highly time consuming and final clearance for usage is given only after successful static firing of rocket motors. No failure or major change in ballistics decides the usage or rejection of rocket motor. A probabilistic approach is also adopted for predicting service life, where the first step is the determination of the stress–strain interference to determine the most probable failure mode. Transportation, storage, handling, environmental loads *etc.* generally induce stress and strain in the propellant. This causes damage in the propellant, which accumulates with the passage of time. The structural analysis of aged propellant indicates the changes in the safety margin as a function of time. Both real time and predictive assessment, using proper failure criteria may be adopted to predict the service life of rocket motor.

12.4.2   Life Extension

12.4.2.1   *Double Base Propellants*

Based on experience, expertise, infrastructure, a number of arbitrary but logical approaches are adopted by various countries for a life extension exercise. For conventional double base propellants, the time of auto-ignition at a given temperature, the time for production of fumes and stabilizer depletion have been used for life assessment. Some of the other qualitative methods include surveillance test, B&J test, vacuum stability test, Woolwich test and climatic hut trials. The time required for depletion of 50% of the original stabilizer content is taken as the useful life of the DBP. Lately, chromatographic techniques have also been used for aging studies. These techniques enable separation, identification and estimation of the stabilizer including other reaction products in a given formulation quickly. HPLC is best suited for this work, although highly nitrated derivatives of stabilizers can be separated by gas chromatography (GC). DTA, TG and DSC provide vital information regarding the deflagration temperature.

Enough literature is available on the aging behavior of double base propellants. It is a well established fact that the rate of deterioration of the propellant increases about 1.8 times with every 10 °F increase in temperature. Life at room temperature = period of test × 1.8^Y, where Y is a multiple of 10 of difference in temperature of test and room temperature in °F.

The Woolwich formula, which is based on depletion of the stabilizer content has been modified and the new formula suggested is as follows:

$$\text{Half life} = T\log 2/\{\log(A/[A-x])\}$$

where 'T' is time for storage, 'A' is initial stabilizer percentage and 'x' is fall in stabilizer percentage.

12.4.2.2 Composite Propellants

The aging behavior of CP are quite different from those of DBP. In the case of DBP, ballistic properties are adversely affected on aging due to migration of the plasticizers, particularly NG towards the inhibitor of the propellant, whereas mechanical properties, which are comparatively much higher (TS 100–140 kg cm^{-2}) are least affected in the case of DBP. In case of CP, both ballistics and mechanical properties may be affected simultaneously. Generally, the propellant becomes harder immediately after manufacture due to post-curing reactions. Moreover, the oxidative species generated from the work horse oxidizer AP decomposition also affect binder degradation.

To determine the mechanical properties deterioration on aging, a commonly used uni-axial tensile test is not considered adequate for structural analysis, as the storage conditions may lead to the development of multi-axial stress fields. Thus, a variety of dynamic tests are needed to measure propellant deterioration to cyclic loading. It has been reported that composite propellants loose their viscoelastic properties with the passage of time and therefore, a relationship must be found between the aging time and the temperature. It has been reported that the Arrhenius equation can be applied for the determination of activation energy (E) of the accelerated aging process:

$$E = 4.572(\log K_2 - \log K_1)T_1T_2/T_2 - T_1$$

where, K_1 and K_2 are the decomposition rates at temperatures T_1 and T_2, respectively.

The 'E' values are extrapolated to the desired storage temperature. This approach requires storage of propellants at constant temperature and determination of parameters like weight loss, burning rate and mechanical properties. A more accurate value of 'E' can be obtained by plotting the log of rate of change of any specified property (h) as a reciprocal of storage temperature T, considering ageing to be a zero order reaction.

$$\Delta \log \mathrm{d}h/\mathrm{d}t = (-E/2.303) - (1/T)$$

12.4.2.3   CMDB Propellants

The time of auto-ignition (TAI) is the most realistic method for the determination of life of CMDB propellants. This involves heating the sample under isothermal conditions in a closed cell of standard volume. This ensures that all the decomposition products are in contact with the propellant sample, thus simulating the worst case conditions under actual propellant aging. TAI is measured by thermocouple at intervals between 100–140 °C and then extrapolated to lower temperature. AP acts as catalyst in the de-composition of CMDB propellants. The determination of rate of depletion of resorcinol, which is used as a stabilizer, is an effective method for determination of life of CMDB propellants. Life of about 50 years is reported for CMDB propellants containing resorcinol as the stabilizer.

The aging properties of CMDB propellants have also been studied after storage at elevated temperatures and determining the change of mechanical properties like TS, relaxation modulus and percent elongation. Resorcinol has been found to be most effective stabilizer. However, combination of two stabilizers namely, resorcinol and two NDPA produces a synergetic effect. Various dihyric and trihyric phenols work as stabilizers for CMDB propellants and produce higher stability. A shelf life of 57 years was obtained for phloroglucinol-based CMDB propellants, which is lower than the resorcinol containing CMDB propellant having a life of 73 years. RDX and PETN based propellant compositions produce higher stability than AP-RDX based formulations. The shelf life calculated from auto-ignition tests has been found to be higher for the RDX based formulation than that of AP and AP-RDX containing compositions.

12.4.3   Life Extension of Propulsion Systems

From an economic angle, the residual life prediction or serviceability of aged propellants/motors assumes high significance. Based on past experience and successful usage of aged propellants/rocket motors, the following methodology is recommended for the life extension of the aged propulsion system of aged missiles.

12.4.3.1 Chemical Composition

The chemical degradation of the propellant can be assessed by qualitative and quantitative analysis of the propellant ingredients, NC, NG, stabilizer, plasticizer, ballistic modifiers for double base propellants and AP/RDX, Al, binder and additives for composite propellants adopting wet analysis and instrumental techniques (IR, UV, HPLC), determination of calorimetric value of the composition using Julius Peter apparatus, and stability tests like B&J test and vacuum stability test. Since the test methods are well documented in the literature for analysis, the details can be obtained from the original source.

12.4.3.2 Radiographic Analysis

The propulsion unit, after aging is radiographed at-least in two orientations, covering the entire length of the motor. During the course of aging, void, gas and crack formation are observed in the propellant, which increases in size with the passage of time. The crack growth propagation of the propellant is a highly complex phenomenon. There are no theories to explain all the observed facts satisfactory. Radio-graphically cleared propellant grain or rocket motor is then taken for static evaluation (rocket test bed). However, propellant charges having major and critical flaws stand rejected. The radiographic acceptance criteria are decided during the final stage of development of a rocket motor and critical/major defects are also defined in the beginning. The basic logic of deciding these defects includes change in surface area due to flaws and their effect on spin pressure inside the rocket chamber and ballistic parameters changes due to observed flaws/defects. Generally, radiographic defects fall in three categories, *i.e.* minor, major and critical. The critical defects include crack formation and debond of the propellant from the rocket case or separation of inhibitor from the propellant charge. Based on the nature and magnitude of defects, a judicious decision is taken for static evaluation.

12.4.3.3 Mechanical Properties Determination

It is possible to assess the physical changes of the propellant by determining the mechanical properties at three temperatures namely, ambient (+25 °C), cold (−40 °C) and hot (+55 °C). The propellant samples are cut and machined from the aged propellants. The samples for these tests are prepared as per ASTM standard. These tests provide quantitative values of compressive strength, percentage compression, modulus of compression, tensile strength, percentage elongation and Young's modulus, which indicate the change in mechanical properties during ageing.

12.4.3.4   Ballistic Performance Test

For measurements of change in ballistic parameters like burn rates, maximum pressure, ignition delay, ignition peak pressure, characteristic velocity *etc.*, a static evaluation of the rocket motor is carried out at cold (-40 °C) and hot ($+55$ °C) conditions. The ballistic properties obtained are compared with the originally specified values of acceptance criteria at the time of development. If the pressure–time and thrust–time profiles match closely with the stipulated values, the life is extended for a specified period and tests are repeated again after 2–3 years.

After critically analyzing the results of chemical composition, density, calorimetric value, mechanical properties, radio-graphic analysis, ballistic properties *etc.*, a decision is taken for life extension or serviceability of the propellant/rocket motor.

12.5   Catalyzed and Platonized Double Base Propellants

The burning rate-pressure relationship is one of the major criteria in the selection of propellants for specific missions. Catalyzed/platonized double base propellants (DBP) offer numerous advantages over conventional propellants (Figure 12.6). Ultra high burn rate (UHBR) propellants increase the thrust of rocket motors and end burning (cigarette) propellant charges may be used in place of perforated charges. Platonized and mesonic propellants capable of producing a pressure exponent (n) value of zero and negative burn rate, $r = aP_c^n$, (P_c is the chamber pressure; a and n, are constants dependent on propellant composition) are not only pressure insensitive but also temperature insensitive, resulting in consistent ballistics at extremes of temperature (-40 °C to $+55$ °C), thereby implying that rocket motors with lower safety factor could be used. For tropical countries like India, where the same rocket is expected to produce consistent ballistics (pressure–time and thrust–time profiles) under adverse temperature considerations, R&D work on these propellants assumes greater significance.

A large number of studies have been conducted in the past in an attempt to find out the most effective catalyst and platonizer in

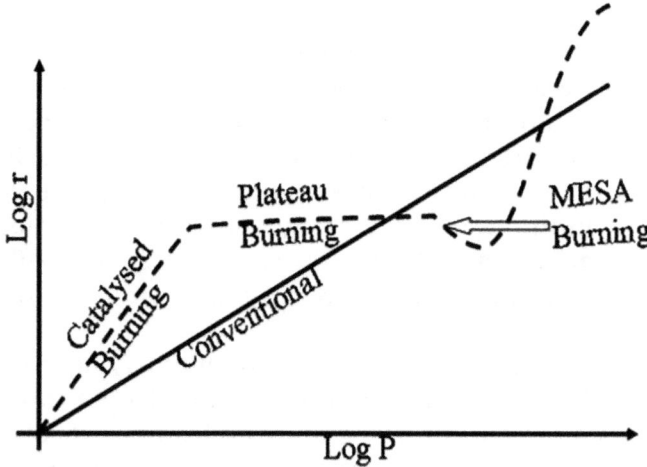

Figure 12.6 The burning rate–pressure relationship.

different pressure regions and to explain the site and mode of action of these additives to enable researchers to choose them in a scientific manner rather than on a trial and error basis involving a large number of experiments. Information on salts capable of producing catalysis/platonization have been published profoundly and a broad summary applicable for DBP have been reproduced here.

Lead and copper salts of inorganic and organic (both aliphatic and aromatic) produce a super rate and plateau behavior. The combination of lead and copper organic salts produce synergistic effect. Other metal salts reported to be effective as a catalyst/platonizer are cobalt, barium, zinc and tin. However, only lead salts have been found to be the most effective as platonizers. Lead salts of aromatic acids produce higher catalytic and plateau effects than lead salts of aliphatic acids. Ballistic modifier highly effective as platonizer for cooler DBP compositions (calorimetric value: 900 cal g^{-1}) act as an 'n' value reducer ($n = 0.3$) for hotter propellant compositions (calorimetric value: 1200 cal g^{-1}). Lead salts decompose in an exothermic mode in the temperature of foam/fizz zones. Catalyzed propellants decompose at lower temperatures and percent decomposition and rate constants are higher, whereas activation energy is lower than the unanalyzed propellants.

The catalytic effect increases with increasing carbon chain length of the lead salts and a shift in the plateau to a higher pressure range is obtained with an increase in carbon chain length of the acids. The inclusion of carbon black along with lead salts further enhances the burning rates and shifts plateau to the higher pressure range. The best results are obtained with a lead to carbon black ratio of 2–2.5. A number of processing parameters, source and purity of starting materials *etc.* influence catalytic/plateau behavior of DBP. Prominent among these are nature and type of plasticizer, origin of nitrocellulose (cotton waste/cotton linters), method of processing (extrusion/casting) *etc.* The minimum number of passes during rolling of extruded DBP is needed. In addition, purity, size and particle distribution of ballistic modifiers and other additives like carbon black *etc.* influences the reproducibility of the pressure exponent and temperature sensitivity coefficient $(\pi_r)_p$ values.

Since lead salts have remained in the limelight as superior catalysts/platonizers relegating copper and other metal salts to a secondary status, research has concentrated more on lead-based compounds.

Lead salt inclusion, however, produces an environmental pollution problem. For realizing ecofriendly catalyzed and platonized propellants other metal salts of mainly cobalt/barium have been studied extensively. Among barium salts, barium β-resorcylate was found to be highly effective platonizer ($n = 0.15$). Cobalt β-resorcylate proved to be a powerful catalyst and platonizer, followed by cobalt salicylate. The surface temperature (T_s) for modified propellants (containing cobalt salts) was higher (460–520 °C) than the control composition (435 °C).

Most of theories proposed to explain the mechanism of catalysis can be broadly classified as:

- A photochemical sub-surface reaction induced by UV radiation from the foam zone.
- An acceleration of NO reduction by carbon and increased carbon formation in the condensed phase.
- An enhancement in the oxidation of the decomposition products of propellant by NO_2.

These theories are based on the kinetics and thermo-chemical analysis of DBP. Catalysts increase the burning rate by enhancing the temperature gradient in the fizz zone. An X-ray photo electron spectroscopy (XPS) study of the condensed phase combustion of DBP with and without lead indicated that the maximum catalysis takes place about 10 μm below the burning surface and that lead promotes the formation of soot in the condensed phase reaction, which accelerates the catalytic oxidizing reaction. The decrease of the burning rate at the plateau/mesa region is caused by the rapid loss of lead from the soot. The plateau effect is attributed by some researchers to the disappearance of carbon frameworks as a result of accelerated oxidation of C by NO and an agglomeration of the catalyst particles thereafter, leading to a reduction of the specific surface and a fall in catalytic effect. Lead contribution to catalysis can be explained by three steps (i) accelerated rate of NO_2 adsorption on lead oxides, which enables the redox reactions between NO_2 and decomposition products, resulting in higher number and density of carbon framework; (ii) holding of catalyst particles by carbon framework and prevent subsequent agglomeration; and (iii) accelerated oxidation of C by NO, which leads to depletion of carbon framework in plateau region. In fact, the carbon produced during burning acts as a catalyst carrier and their high activity makes the reaction species easily adsorbed. The oxidation and disappearing rate of the carbon carrier is affected by the surface activity of carbon black. As per another proposed mechanism of catalytic action, a key event is interception of NO_2 by lead and prevention of the reversal of the nitrate ester dissociation. Lead intrudes into the decomposition cage of nitrate ester and provides a low energy path for NO_2 escape. In this mechanism, lead acts as a renewable reagent rather than a catalyst and interaction of lead with NO_2 lowers the activation energy of nitrate ester decomposition.

For catalysis, the luminous gas phase reaction does not contribute in a significant way but foam and fizz zone reactions are influenced by metal oxides and metal salts. The formation of a new carbon network in the catalyzed propellants is a major support to catalysis. It is well known that the reduction of NO by aldehydes, CO, H_2 or carbon is a highly exothermic process and in lead catalysis the formation of CO_2 and carbon is observed. It is expected that any metal oxide or metal salt capable of forming higher carbon species catalyzing $NO_2 + CH_2O$ and NO + C reactions should be an effective catalyst/platonizer.

It appears catalysis and platonization depend on the C/NO ratio in the fizz zone and in general if C/NO is more than 1, a super rate is observed, while at C/NO equal to 1, a plateau effect is observed. Since several possible mechanisms have been reported and the specific effects have not been characterized fully, these phenomena still remain an enigma to solid propellant scientists.

12.6 Advanced Solid Propellants

During recent decades, composite propellants based on AP/Al/HTPB have gained wide acceptance as a source of energy for propulsion of rockets/missiles and space vehicles. In addition to propulsion, propellants are also being used increasingly in oil and gas production, fire fighting and life saving devices and gas generators. Hence, a continuous search is on to develop more powerful propellants, which means higher Isp. A higher Isp value extends the flight range and increases the payload. Furthermore, if the density of the propellant is also higher, a greater value is realized due to the relative reduction of rocket dimensions and rocket case weight. For military applications, the burn rates of propellant should be independent of pressure and temperature. This is possible to achieve with the help of ballistic modifiers. In addition, high mechanical properties of propellant provide high loading of rocket motors.

An ideal propellant for rocket/missiles/space vehicles should have the energy of cryogenics (390 s), density of solids (2 g cm^{-3}) and the ability to tailor ballistic properties like burn rates, pressure index (n), temperature sensitivity coefficient (π_{rp}) and mechanical properties. The energy potential of various solid propellants is given in Table 12.4.

Hybrid propulsion systems (HPS) based on a liquid oxidizer and solid fuel are gaining importance for space vehicles due to their higher safety and attractive energy features. A comparison of potentials for solid, liquid and hybrid propellants are given in Table 12.5.

The various components of advanced solid propellants include energetic oxidizers, energetic binders and energetic additives. A few special purpose additives are added to the propellant to achieve desired properties. Bonding agents and process aids are the other materials needed for advanced propellant processing.

12.6.1   Energetic Oxidizers

AP-based composite propellants introduce noxious materials into the stratosphere. To avoid this limitation, new oxidizers are being investigated. The new global slogan is to develop clean and alternate propellants. This demands waste and hazardous material reduction,

Table 12.4   Energy potential for various solid propellants.

Types of propellants	Specific impulse, Isp (s)
Double base propellants (DBP)	200–220
Nitramine-based DBP	230–250
Composite propellants (CP)	240–260
Composite modified double base (CMDB) propellant	250–270
Fuel-rich propellants (FRP) in IRR mode	200–1200
Advanced solid propellants (ASP)	>300

Table 12.5 Energy levels of solid, liquid and hybrid propellants.

Propellant type	Energy (Isp, s)
Solid	200–300
Liquid	
a. Storable	
i. HNO_3 + UDMH	276
ii. HNO_3 + N_2H_4	285
iii. N_2O_4 + UDMH	285
iv. N_2O_4 + N_2H_4	290
b. Cryogenic	
i. O_2 + H_2	370
ii. F + H_2	390
Hybrid	
(a) IRFNA + HTPB	280
(b) N_2O_4 + HTPB	300
(c) H_2O_2 + HTPB	300
(d) F + HTPB	350

Table 12.6 Properties of solid oxidizers.

Oxidizer	Molecular formula	Mp (°C)	ΔH_f (kJ mol^{-1})	ρ (g cm^{-3})	OB (%)
AN	NH_4NO_3	170	−365	1.72	20
AP	NH_4ClO_4	130	−296	1.95	34
HP	$N_2H_5ClO_4$	170	−178	1.94	24
HP2	$N_2H_6(ClO_4)_2$	170	−293	2.20	41
NP	NO_2ClO_4	120	37	2.22	66
RDX	$C_3H_6N_6O_6$	204	71	1.82	−21.6
HMX	$C_4H_8N_8O_8$	278	75	1.96	−21.6
ADN	$NH_4N(NO_2)_2$	90	−150	1.82	25.8
HNF	$N_2H_5C(NO_2)_3$	295	−72	1.90	13.1

ingredient substitution and toxic release prevention. The identification and potential high performance propellant combination is based on a fundamental analysis of the governing parameters. Safety and handling requirements and consideration of thermodynamic decomposition and explosive properties of new materials also assume high importance. The elementary properties of solid oxidizers are given in Table 12.6.

It can be seen that both ADN and HNF appear to be highly promising oxidizers. However, HNF is better than ADN in terms of its melting point, density and heat of formation. The reduction of unwanted HCl can be partly obtained by replacing part of AP by AN and HMX. HCl reduction figures are given in Table 12.7.

12.6.2 Energetic Binders and Plasticizers

In the field of energetic binders and plasticizers, a range of compounds that contain azido (N_3), nitro (NO_2) and nitrato ($O-NO_2$) or difluro amino groups are preferred. Prominent among the energetic group is the azido group, which contributes about 355 kJ per N_3. The first energetic polymer used was GAP in 1990. Subsequent research led to polyglycidyl nitrate (PGN), nitromethyl methyl oxetane (NIMMO) as promising compounds. The physico-chemical properties of GAP, NIMMO and BAMO are given in Tables 12.8 and 12.9.

Among the plasticizers, nitro and nitrato compounds are gaining importance. Trimethylol ethane trinitrate (TMETN), bis dinitro propyl formal/acetal (BDNPF/A) is another plasticizer system of practical utility. Low molecular weight GAP (mol weight 400–700) can be used as an energetic plasticizer. It desensitizes NG and is therefore very attractive for energetic double base propellants. Poly-BAMO is prepared by the chlorination of pentaerythratol and cyclization of trichloride to bis chloromethyl oxetane (BCMO in first step, followed by reaction with NaN_3). Likewise, AMMO is prepared by the action of NaN_3 on

Table 12.7 HCl reduction on part replacement of AP by AN.

Prop. ingredient	Comp I	Comp II	Comp III	Comp IV	Comp V
AN	—	36	31	41	41
AP	69.7	10	15	10	10
HMX	—	—	—	—	18
Al	16	18	18	18	10
HCl emission	21	2.8	4.2	2.8	3.0

Table 12.8 Properties of energetic binders.

Properties	GAP	P-AMMO	P-BAMO	P-NIMMO
Color	Pale yellow	Yellow	White solid	Yellow
Mol Wt	700–5000	3000–4000	2000–2500	2000
Viscosity (cps)	1200 (20 °C)	—	—	1600
T_g (°C)	−55	−35	−45	−30
ΔH_f (kJ mol^{-1})	117	345	246	334
Thermal stability	> 200 °C	—	—	—

chloromethyl emthyl oxetane (CMMO). Plasticizers are added to binders to realize the desired level of process ability. Azido plasticizers have been extensively studied as energetic plasticizers. Poly-NIMMO based gas generators have stable burn rates at low pressure (5 MPa) with reasonable pressure index values.

Nitro-esters such as EGDN and NG are of much interest to the propellant community. Aliphatic nitramines like diazido 3-nitra aza pentane are very promising plasticizers for energetic gun propellants. Energetic additives, which can enhance specific impulse are given in Table 12.10 along with their properties.

Table 12.9 Properties of energetic binders.

Name	Structure	Density (g cm^{-3})	Impact sensitivity (ksc)
GAP	CH_2N_3 $-(CH_2-CH-O)_n$	1.3	>170
Poly-Glyn	ONO_2 $-(CH_2-CH-O)_n$	1.42	>200
Poly-NIMMO	H_3C CH_2NO_2 $-(O-CH_2-C-CH_2)_n$	1.26	>90
Poly-AMMO	H_3C CH_2N_3 $-(O-CH_2-C-CH_2)_n$	1.06	>90
Poly-BAMO	N_3H_2C CH_2N_3 $-(O-CH_2-C-CH_2)_n$	1.3	>200

Table 12.10 High energy additives.

Name	Formula	Density (g cm^{-3})	DP (GPa)	VOD (m s^{-1})
Cl-20 (HNIW)	$C_6H_{12}N_{12}O_{12}$	2.10	42	9400
Octanitro cubane	$C_8N_8O_{16}$	2.00	47	9800
High nitrogen compounds	N_8	2.65	137	15000
High nitrogen compounds	N_{60}	2.67	196	17300
Poly-nitro adamentane	$C_4H_4 N_{12}O_{12}$	2.10	43	9500

With regards to the performance of advanced solid propellants, the higher the solid loading, the higher the performance. Smoke gets considerably reduced by the inclusion of nitramines (RDX/HMX). However, ecofriendly combustion products along with higher energy can be obtained by inclusion of powerful oxidizers like ADN and HNF. It attacks the double bond of butadienes. The HNF-GAP-Al combination appears to be very promising from an energy point-of-view. Hence, HNF–Al_GAP based compositions have been identified as very promising propellants. The incorporation of metallic hydrides (LiAlH$_4$) can still enhance energy (Isp). Zirconium having a high density (6.5 g cm^{-3}) is a very promising candidate for high volume Isp systems. The predicted performance of advanced propellant systems is given in Table 12.11.

12.7   Nanomaterials in Rocket Propellants

Nano (10^{-9}) represents extremely small physical dimensions, which are not perceived by the naked eye and any material that has at least one of its dimensions in nanometers is classed as a nanomaterial. The revolution or charm of nanotechnology has engulfed almost all domains of modern production, manufacturing, design, synthesis, characterization and industry. A material, which is a conductor becomes a non-conductor at the nano levels, a yellow material becomes red at the nano level, a ductile material behaves in a brittle manner in the nano state, and so on. As science and technology has to be redefined in nanometric sizes, afresh, the role of nanomaterials as ingredients in solid propellants for rockets must be explored. Nanomaterials are characterized by high chemical reactivity, large surface-to-volume

Table 12.11   Predicted performance of advanced propellant systems.

Propellant system	Isp (s)	$T_f(^\circ C)$
AP-Al-HTPB	250	2742
HNF-AL-GAP (12)	271	3092
ADN-Al-GAP (18)	262	2838

	Binder			
Oxidizer	BAMO	GAP	PLN	PGN
ADN	306	304	303	300
HNF	311	308	307	304

ratios and alternate combustion response. The scope for performance improvement by incorporation of nanometric ingredients in solid propellants formulations may be considered apt for developing modern propellants. Many researchers have explored use of nanomaterials in solid rocket propellants and various effects are reported. This part of the book covers in brief, the results, available in open literature, where consensus in improvement of performance could not be established by incorporation of nano ingredients.

Since conventional double base propellants based on NC and NG have very little scope for solid ingredient incorporation, they are not affected much by the advent of nanomaterials. In addition the lower quantity of solid ingredients, the well-established production plant and advent of more energetic and promising composite propellants makes double base propellants virtually immune to the nanomaterials revolution. Composite propellants have around 80–85% solid ingredients in the form of oxidizers (65–85%) and metallic fuel (10–20%). In addition to this, burning rate modifiers (1–5%), as transition metal oxides are also particulate matters present in the composite propellant formulations. The workhorse oxidizer is AP and Al is used as the metallic fuel in the composite propellants. Iron oxides, copper chromites *etc.* are used as solid burning rate modifiers in the composite propellants. This section deliberates the effect of these ingredients in nanosized proportions and also some new propellant formulations, which are not possible, otherwise at the micron or higher particle size ingredients.

Since AP is around 65–85% in the composite propellant formulation, and its decomposition affects the burning rate of the propellants, decomposition of AP is investigated by many researchers, exhaustively. Transition metal compounds are known to catalyze the burning behavior of AP. A specific feature of the thermal decomposition of AP is the influence of additives on high temperature thermal decomposition of AP and deflagration delay time. DSC curves for pure AP shows an endothermic peak at 243 °C due to the crystallographic transition from orthorhombic phase to cubic form. This transition is generally not altered by any additives, be it metals or any transition metal oxides. The first low temperature exothermic peak is observed at 331 °C, which indicates the partial decomposition of AP. The second and main exothermic peak occurs at 467 °C, indicating complete decomposition of AP. The exothermic peaks are altered by the presence of metals and burning rate modifiers. TGA of AP in the micron and nanosized dimensions are different. In the TGA of micron sized AP, there is a smooth loss of weight with an increase in temperature.

However, the TGA of 35 nm-sized AP, gives a sharp two stage weight loss. 20% weight loss was observed in the temperature range of 310–350 °C, while complete decomposition was observed around 400 °C. Complete decomposition of nanosized AP occurs at a lower temperature than micron-sized AP. The agglomeration of nanometric AP is an inherent drawback and the same is always observed by XRD. The replacement of micron sized AP with nanosized AP in AP/Al based propellant formulation, the burning rate increases from 2.5 mm s^{-1} to 13.5 mm s^{-1}.

During thermal analysis, oxidation of Al is mainly observed and this leads to an enhancement in the weight. As far as Al powder is concerned, the major part of 97 micron-sized Al oxidizes in the temperature range of 740–1000 °C and they remained in the alpha-phase. For 10–50 nm-sized Al, as characterized by AFM, 3.5 times the weight gain was observed in the temperature range of 500–740 °C. This is due to formation of aluminium oxide which remains in the gamma-phase. The lowering of oxidation temperature is responsible for a change of phase for nanosized Al-particles. In the AP/Al-composition, when micron sized Al was replaced with nanosized Al, an increase in burning rate from 2.5 mm s^{-1} to 26.5 mm s^{-1} is observed.

The role of Al in the composite solid propellant is altered by a reduction in its particle sizes. Using nanosized Al with AP can increase the burning rate of dry-pressed pellets and the decomposition temperature of AP is also found to reduce by the presence of nanosized Al. However, the low pressure deflagration limit (LPDL) is enhanced with decreasing particle size of Al and increasing Al content. For pure AP, LPDL is 2 MPa, while with 10% micron-sized Al, the value reaches 4 MPa. The addition of 10% nanosized Al raises the LPDL to 12 MPa. In fact, AP-monopropellant flame is weak and it does not possess sufficient heat to pyrolyse or ignite Al. So, nanosized Al particles accumulate at the surface of the propellant and act as a heat sink. This heat sink effect is more pronounced for nanosized Al. The burning rate of AP-pellets with micron-sized Al is higher than that with nanosized Al. This observation is contrary to the effect observed in composite propellant formulation where micron sized Al is replaced by nanosized Al. Micron sized Al does not burn in the AP flame and it is carried away from the AP burning surface. Sometimes the high oxidizer content at the surface of nanosized Al is also attributed to a lower burning rate with nanosized Al. Such inhibition of ignition is possible at low pressures and at high pressure enhancement in burning rates is possible.

The addition of 20% ultrafine (nanosized) Al powder (called ALEX) is reported to increase the burning rate of the propellant (based on

HTPB/AP/Al) by 40% and mass burning rate by 70%. The advantages of using nano Al in propellant formulations are: (i) to shorten ignition delay; (ii) to shorten the burning time and get more complete combustion; (iii) increase the heat released due to dispersion in high temperature zone by direction oxidation; and (iv) enhance the heat absorptivity of the propellant. In further studies, it is observed that despite the adverse condition of reduced energy release per unit volume of ALEX particles, the larger surface area, increased reactivity with gas-phase species, shorter ignition delay and high bulk mass burning rate of ALEX gives an increased burning rate, increased temperature sensitivity of propellant formulation and increased pressure index of burning rate in propellant formulations. Despite higher volume fraction of passive or inert oxidizer in nanosized Al, the mechanism of the burning rate enhancement is explained on the basis of larger surface area (more reactivity in gas-phase), shorter ignition delay and higher mass burning rate of ALEX due to high surface to volume ratio. The burning rate of the composition, containing ALEX increases. It is observed that at low pressure (10 atm), the burning rate enhancement is 2.20 times, while at relatively moderate pressure (60 atm), the burning rate enhancement due to replacement of 6% micron sized Al by ALEX, is only 1.25 times. At further higher pressures, the burning rate advantage due to such replacement is expected to vanish. Complete replacement of micron sized Al with ALEX gives a propellant formulation, which shows unstable combustion at elevated pressures and burning rate is found to be independent of pressure (very low pressure index). When 35% AP is replaced with HMX for a new set of formulations, replacement of 6% micron-sized Al by ALEX enhances burning rate by 50%. In both the compositions, replacement of micron-sized Al by ALEX results in reduction in pressure index.

Identical aluminized AP/HTPB-propellant formulations with ALEX (nanosized Al) and micron sized (15 micron) Al are prepared for combustion studies. The surface of extinguished propellants, with ALEX, is smoother. Smaller holes created by ejection of Al-particles are also visible and relative hole size measurement confirms agglomeration of ALEX particles by melting at the surface of the extinguished propellant surface. The combustion of the propellants with ALEX exhibits higher burning rates without affecting pressure exponent of burning rate, maximum temperature and clearer radiative emission in the gas phase. For reference composite propellant containing HTPB (15%), AP (80 micron size, 60%), RDX (320 micron size, 20%), Al (5%), burning rate data using strand burner are generated in nitrogen pressurized chambers at pressure range from 2 MPa to 10 MPa. The burning rate reduces by part replacement of micron-sized Al by nano Al and surprisingly the burning rate index reduces. However, when a higher quantity of micron-sized Al is replaced with nanosized Al, the burning rate enhances but the burning rate index is found to be lower than the reference composition.

Nanosized Al particles manufactured by an electro-condensation process gives spherical particles of 43 nm with a passivating oxide layer of around 3 nm. Doping of outer surface of nano Al with barium shows a suitable thermal behavior in the oxidation reaction, a considerable burning rate enhancement and the suppression of agglomeration in solid rocket propellant formulations. The replacement of micron sized Al by barium-doped nano Al results in a considerably larger steady burning rate and decreases agglomeration in AP-based composite propellants. Other chemicals like benzene, silicon rubber are also used for doping, but the pressure build-up rate of barium doped nano Al is around four times faster than nano Al doped with other chemicals. Compared to micron size Al, the burning rate improves 10 times with nanosized Al. The size of the agglomeration with micron sized Al is 28 microns, which becomes only 2 microns for nano Al. This results in reduced slag formation, low two-phase flow losses and enhanced efficiency of combustion. The XPS spectra detects the outer surface of the particle as composed of Ba^{2+} ions bonded to BaO and Al^{3+} ions bonded to Al_2O_3.

The Use of nano Ni was also investigated and with 2% nano Ni added by parts, the burning rate improves and the burning rate index reduces. Without nanosized Al or with less nanosized Al, the flame has larger agglomerates away from the surface, but a large nanosized Al percentage in the composition or use of nano Ni reduced size of agglomerates in the flame and brought it near the burning surface. Two parts of iron oxide 3.5 nm are added in the composite propellant formulation containing 85% AP (oxidizer) and 15% HTPB (binder) and DSC studies are conducted. High heat release, increased rate constant and lowering of high temperature decomposition confirms enhancement in the catalytic activities of AP in the presence of nano iron oxide. The increase in activation to 181.5 kJ mol^{-1} (from 143.8 kJ mol^{-1} for base composition without nano iron oxide) is due to the addition of nano iron oxide.

Ultrafine Al particles are incorporated in AP based different solid rocket propellants based on two different binders PBAN and PCP. Although an increase in the burning rate due to reduction in particle size of Al is observed, a complicated surface concentration with sintering-cum-agglomeration is stated to be responsible for such deviations.

Although for micron sized Al particles, PCP based propellants gives higher burning rates than PBAN based formulations probably due to leading edge flames (LEF), but the trend reverses with nanosized Al particles. The probable reason may be impeded release of heat near the propellant surface, reduced luminosity of the Al-burning region and PCP encouraging agglomeration. If all micron-sized Al is replaced with ultrafine Al, excessive agglomeration causes reduction in heat release near the propellant surfaces.

Ballistic modifiers like iron oxide, copper chromite, ferrocene-derivates *etc.* are used in composite propellants for the enhancement of burning rates. Similarly, oxamide, lithium fluoride *etc.* are burning rate suppressants for composite propellants, lead and copper salts are used as catalyzing and platonizing agents in double base propellants. These ingredients in micron sizes are established to provide this ballistic or burning rate modification effects. But nanometric forms of these ingredients may be behaving differently in solid propellants. The addition of micron sized copper chromite ($CuCr_2O_4$) and cupric oxide (CuO) alters both the exothermic peaks. It reduces the high temperature exotherm, significantly. However, the effect of nanosized particles is more pronounced. With nanosized cupric oxide, high heating rates lead to an overlap of the decomposition at both temperatures. The high temperature exotherm shifts from 467 °C for pure AP to 360 °C for 2% w/w CuO at a heating rate of 10 °C per minute. The addition of 2% w/w $CuCr_2O_4$ gives a high temperature exotherm at 349 °C at the same heating rate. The heat release by addition of these nanosized additives to AP leads to a higher value of heat release also. For pure AP, the heat release is 0.834 kJ g^{-1}, while with 2% w/w CuO, the heat release becomes 5.430 kJ g^{-1}. This may be attributed to high surface area of nano CuO, which prevents sputtering of particles during decomposition, reduces mechanical loss and gives more efficient heat transfer.

Copper/carbon nanotube particles, prepared by a precipitation method, demonstrate excellent catalytic performance, on the thermal decomposition of AP. The high temperature decomposition peak of AP is reduced by 126.3 °C and the low temperature decomposition peak disappears. The addition of these materials, as simple mixtures, reduces the high temperature decomposition peak of AP by 114.9 °C. Metal nanopowders are used for reducing thermal decomposition temperature of AP and onset of decomposition at 307 °C for pure AP is found to reduce to 244 °C, 291 °C, and 146 °C for addition of iron, nickel and copper nanopowders, respectively. Thermal decomposition of AP is affected by transition metal oxides and that is why iron oxide and copper chromite were good burning rate modifiers for com-position solid propellants based on AP. Nanocrystalline transition metal oxides (Cr_2O_3, Fe_2O_3, Mn_2O_3), prepared by three different meth-ods (novel quick precipitation method, surfactant mediated method, reduction of metal complexes with hydrazine as reducing agent), are used for thermal decomposition studies of AP. These transition metal oxides shifts high temperature decomposition of AP to lower temperature. Binary transition metal ferrite (BTMF) nanocrystals of general formula MFe_2O_4 (M = Cu, Co, Ni) are prepared by a co-pre-cipitation method and characterized by XRD. The particle sizes are confirmed to be in the range of 30–50 nm. The BTMF shifts the high temperature decomposition stage of AP towards a lower temperature and the sequence of decreasing effectiveness is found to be cobalt, copper and nickel.

Hydrazinium nitroformate (HNF) is an alternate energetic oxidizer, whose high sensitivity to friction is a major obstruction to its use. The reduction in sensitivity of HNF is attempted by improving purity, modifying crystal morphology, mechanical crystallization, sono-crys-tallization, phlegmatization, drowning out, co-crystallization and crystal shape modifiers. However, only limited success in friction sensitivity reduction is achieved by these methods. Hence, a multi-pronged approach is adopted by researchers. Using sono-mechani-cal crystallization with crystal shape modifiers, followed by double coating of HTPB based nano-composite decreased sensitivity of HNF from 2.0 kg to 9.6 kg. The impact sensitivity was also increased from 25 cm to 55 cm by this approach. With 5% by weight of nanocompos-ite, the friction and impact sensitivity reduced to 6.4 kg and 36 cm, respectively. With 2.5% by weight of crystal shape modifier and 5% by weight of single coated nanocomposite, the values changed to 7.2 kg and 44 cm, respectively.

Nanosized particles open up a new domain of propellant formula-tions. Al with polytetrafluoroethylene (PTFE) formed nanoenergetic composites, which properties similar to propellants. Ignition, combustion characteristics, pressurization rates and total impulse are evaluated. Since such compositions with micron sized Al powder may not be producing significant energetic effects, a comparison of nanoenergetic composites is not made with corresponding micro-en-ergetic composites. A comparison of performance of nano Al indi-cates that higher burning rates did not necessarily give high total impulse from such nanoenergetic compounds. One of the difficulties to use nanosized ingredients in high energy materials formulations is stated as homogeneous and uniform dispersion of such particles in the entire volume. As an initial study dispersion of nano Al in binders consisting of sodium carboxymethylcellulose mixed with deionized water, in a twin-screw extrusion is studied. The scale of mixing and type of mixer affects the degree of mixedness, dispersion and agglomeration of nanoparticles. The use of specialized equipment is also advocated. A uniform dispersion and reduced agglomeration is promoted by incorporating surfactants.

After reviewing the incorporation and use of nanosized ingredients in propellant formulations, some improvements in the product performance (mainly burning rate and burning rate index) is practically observed. The decomposition of AP is greatly altered by not only nanosizing AP, but also by incorporation of nanosized Al powder and ballistic modifiers. Al powder, which is incorporated as a combustion instability suppressant, density enhancer can act as burning rate modifier in nanoforms. The nanosized burning rate modifiers and certain new propellant formulations are studied by researchers with marked improvement in burning rates. However, the enhancement in energy in terms of Isp is never claimed by the use of nanosized ingredients. The problems with nanosized ingredients have three concerns: cost, passivation, hazard. This collectively limits the usages of nanosized ingredients in propellant formulations. The practical utility for operational systems with nanosized ingredients in propellants is still a distant reality.

12.8   Futuristic Rocket Propellants and Propulsion Systems

Rockets in the future are expected to deliver greater ranges, higher velocities and high altitudes by employing a propulsion system with a smaller size, and lighter and higher density energetic propellants. The improvement in various components of the propulsion system is needed for high performance, ecofriendly exhaust products, and safe-reliable system operations. It needs a reduction in potential leak paths, failure modes and failure causes. The current direction concentrates on effective heat resistant seal materials, Stronger steel case materials, fewer internal nozzles, better mechanical case joints, improved ignition systems and highly automated, reproducible manufacturing and inspection technique.

The improvement in processing technique has several facets. Batch mixing of propellant slurry can be replaced by continuous mixing. This will have advantages in case of filling large size boosters and better reproducibility, least exposure of propellant to human beings, less mechanical actions, reduced propellant waste and better monitoring. Soluble core or collapsible core techniques can significantly enhance safety and potential damage to the propellant during the course of processing.

During last three decades, propellants have been used in −30 °C to +55 °C temperature range, except some countries like Russia, where operation range was from −40 °C to +40 °C. However, due to extension of flight domain, particularly altitude and speed of aircrafts, the working range may change to −55 °C to +70 °C for air-launched missiles. This will demand high mechanical properties of propellants. Other parameters, which may become important from an application point-of-view, are geometry of propellant grain and propellant grain-motor case interaction with respect to deformation under pressure and due to thermal expansion characterization.

For the next generation of propulsion systems, four major requirements will include higher energy (Isp), superior mechanical properties, ecofriendly combustion products and low vulnerability. From economic considerations, QC tests and NDT will become more significant for futuristic propulsion. In addition to chemical propulsion, ion and electrical propulsion/nuclear propulsion may get higher attention. For insulation materials, improved ablation rate and reduced density are governing criteria for future innovations. High melting point fillers reinforced with carbon fibers are being tried. EPDM (Ethylene Propylene Diene Monomer) polymer, micro-balloons and low density carbon filler could cut density by 10–15%. Also in line is the development of smokeless insulation. To improve processing, sheet extension mechanism, vacuum installation, sling or rotary application and IR or other fast curing methods can make current manual and slow curing operation faster. In fact, propellants with low burning rates can be used as thermal insulation layers for producing extra energy at the same time protecting the casing from high heat flux of combustion gases.

The prediction for solid propellant performance is another concern, which must be addressed. The conventional steady state bulk-calculation of internal ballistic parameters is no-longer effective. Erosive burning, combustion instability, aero-thermo-chemical variation in flow field, unconventional ignition pattern *etc.* are some areas, which require better computation tools. If combustion of a solid propellant inside rocket motor casing is closely observed, three areas for exploration are clearly visible: flow pattern evolution, combustion of propellant and incompressible fluid–structure interaction. In an idealized calculation, flow is assumed to be a one-dimensional single phase steady flow, but in reality, turbulence, two-phase flow, alumina droplet formation, distributed combustion, reaction in gaseous products, *etc.* change flow pattern completely. The propellant is assumed to burn in layers, but unsteady combustion, lifted combustion, melt-layer formation, heat sink effects, flow controlled combustion, pressure and temperature, thermal conductivity of product gases, heat transfer coefficient in the vicinity of the solid surface alters combustion variables, significantly. The structure of the rocket includes all solid parts like the motor, the propellant grain, the propellant holding grid, the de-laval nozzle. These structures may be vibrating or absorbing the vibrations created by gaseous flow. The study of these needs proper attention in the future and the future of chemical propulsion is not limited to ballisticians, alone, but structural analysis, fluid dynamics, acoustics, computational expertise *etc.* must be incorporated in prediction strategies.

Although chemical propulsion has remained at the forefront of rocketry, other propulsion concepts are being researched to generate competition. Most of such propulsion concepts are peeping for an introduction in the space exploration arena. Many of them have been tried as auxiliary propulsion systems in spacecrafts. ION propulsion, which is deliberated in brief in Chapter 1 has been used for auxiliary propulsion in the Deep Space I mission (launched in 1998). Aeronautical engineers believe that electrical and ion propulsion will be widely utilized in the 21st century, especially for deep space missions. An ion propulsion system has three components: (i) a xenon propellant source (ii) an electrical power processor, and (iii) a cylindrically shaped thruster. Though ION propulsion typically generates thrust in milli Newtons, the force suffices the tasks it performs. Moreover, this type of propulsion can generate speeds of 25 000 m s^{-1}, which chemical-based propulsion cannot reach. In addition, unlike chemical propulsion systems, ion rockets operate for months and even years. Thus, this propulsion technique is one of the most promising for use in deep space exploration. Additional methods include solar and laser thermal propulsion: essentially chemical propulsion systems that use solar or laser light to heat a hydrogen fluid. In order to heat up the fluid, sunlight must be directed towards these metallic, balloon-like inflatable mirrors. These mirrors, in turn, are meticulously pointed through a small window to the engine, where the propellant is located.

Propulsion without fuel is another concept being explored through Tether propulsion. It is one of the most favorable methods since it uses no fuel, is recyclable and is environmentally safe, and best of all, reduces the cost of spacecraft mobilization to only several hundreds of dollars per pound as opposed to thousands of dollars. Spacecrafts implemented with this system essentially utilize a simple 5-kilometer-long aluminum wire tether as a basis for propulsion. As the tether moves across the Earth's magnetic field, a voltage is induced across its length. The upper end of the tether becomes positive, so electrons in the upper-atmosphere plasma, which carry negative charge, are attracted to the upper end. By using a device to emit the electrons back into space at the lower end of the tether, a current can be made to flow along the length of the wire. This movement of electrons gives thrust or in others words, acceleration, which in turn allows the spacecraft to move. In testing this propulsion system, NASA experimented with the Space Shuttle but failed when the wire broke. Though this mode of propulsion is expected to bring benefits, much work lies in wait for NASA engineers.

Furthermore, there are theoretical propulsion systems that would be revolutionary to say the least, if ever realized. For instance, scientists are investigating nuclear fission and fusion that offer 100 times more energy compared to ordinary chemical propulsion systems. Better yet, matter–antimatter annihilation is viewed to be the ultimate source with a theoretical energy density of 1000 times of conventional chemical propellants. Such mind-boggling possibilities exist since the exhaust velocities of these theoretical systems are extremely high. Chemical propulsion systems are limited to a velocity of approximately 10 km s^{-1}. Those that utilize plasma as a source of heat can move within 20 km s^{-1} to 50 km s^{-1}. Ion propulsion systems, however, surpass both types, having a potential velocity of 200 km s^{-1} or more. Theoretical propulsion concepts like matter–antimatter annihilation drastically transcend all types of propulsion methods currently being studied. Matter–antimatter annihilation, for instance, can theoretically achieve an exhaust velocity comparable to the speed of light, 300 000 km s^{-1}.

Reviewing the past and contemplating the current rate of technological advance, it can be concluded that the boundaries of our potential are endless. There are a few companies that have invested in this commodity, offering to the public an opportunity to take a vacation outside Earth. One ambitious travel agency, Zegrahm Space Voyages, is presently taking reservations for the first tourist trips into space. For a cost of $98 000, one can spend some time in space and view the Earth from an entirely different perspective. It is only a matter of time.

Bibliography

For further reading refer to the following literature

1. N. Kubota, *Propellant and Explosives: Thermochemical Aspects of Combustion*, Wiley-VCH GmbH, Weinheim, Germany, 3rd edn, 2015, ISBN – 978-3-527-33178-9.
2. G. P. Sutton and O. Biblarz, *Elements of Rocket Propellants*, John Wiley and sons, USA, 8th edn, 2010, ISBN – 978-0-470-08024-5.
3. H. S. Mukunda, *Understanding Aerospace Chemical Propulsion*, Interline Publishing, 2004.
4. R. Meyer, J. Kohler and A. Homburg, *Explosives*, Wiley-VCH, 5th edn, 2002.
5. G. D. Roy, *Advances in Chemical Propulsion Science to Technology*, CRC Press, Florida, USA, 2002.
6. E. L. Fleeman, *Tactical Missiles Design*, AIAA Education series, United States, 2001.
7. M. J. L. Turner, *Rocket and Spacecraft Propulsion, principle practice and new developments*, Praxis publishing, UK, 2000.
8. V. Yang, T. B. Brill and W.-Z. Ren, Solid Propellant Chemistry, combustion and Motor Interior Ballistics, *Progress in Astronautics and aeronautics*, AIAA, 2000, vol. 185.
9. R. Balakrishnan, *Guided Weapon System Design*, DRDO, 1998.
10. P. J. Turchi, *Propulsion Techniques Action and Reaction*, AIAA, Verginia, USA, 1998.
11. S. Krishnan, S. R. Chakravarthy and S. K. Athithan, *Propellants and Explosives Technology*, Allied Publishers Limited, Chennai, India, Dec. 1998.
12. S. Krishnan and S. R. Chakravarthy, *Modelling and performance prediction in rockets and Guns*, Allied Publishers Limited, Chennai, India, Dec. 1998.
13. R. G. Lee, T. K. Garland-collins, D. E. Johnson, E. Archer, C. Sparkes, G. M. Moss and A. W. Mowat, *Guided Weapons*, Brassy's London, 3rd edn, 1997, vol. 5.

14. K. Kishore and K. Sridhara, *Solid Propellant Chemistry – Condensed Phase Behaviour of AP based Solid Propellants*, IISc, B'lore, India, Dec. 1996.

15. G. E. Jenson and D. W. Netzer, Tactical Missile Propulsion, *Progress in Astronautics and Aeronautics*, 1996, vol. 170.

16. P. W. Cooper, *Explosives Engineering*, VCH Publishers, New York, 1996.

17. Advisory Group for Aerospace Research and Development, Service life of Solid Propellant Systems, AGARD-CP-586, *Papers from symposium in Athens*, Greece, 10–14 May 1996, ISBN 92-836-0036-3.

18. C. D. Brown, *Spacecraft Propulsion*, AIAA Education series, Ohio, 1995.

19. A. Davenas, *Solid Rocket Propulsion Technology*, Pergamon Press, UK, 1993.

20. L. DeLuca, E. W. Price and M. Summerfield, Non Steady Burning and Combustion Stability of Solid Propellants, *Progress in Astronautics and Aeronautics*, 1992, vol. 147.

21. J. E. Field and P. Gray, *Energetic Materials*, The Royal Society, London, UK, 1992.

22. T. Urbanski, *Chemistry and technology of explosives*, Pergaman Press Ltd, London, 1990, vol. 1–4.

23. S. N. Bulusu, *Chemistry and Physics of Energetic Materials*, Kluwer Academic Publishers, Neatherlands, 1990.

24. Y. M. Timnat, *Advanced Chemical Rocket Propulsion*, Academic Press, 1987.

25. K. K. Kuo and M. Summerfield, Fundamentals of solid propellant combustion, *Progress in Astronautics and aeronautics*, AIAA Series, New york, 1984, vol. 90.

26. *Encyclopedia of Explosives & related items*, US Army Armament Research & Development Command, 1983, vol. 1–10, PART 2700.

27. S. Fordham, *High Explosives and propellants*, Pergamon press, New York, 1980.

28. T. L. Davis, *The chemistry of Powder and explosives*, 1975.

29. R. W. James, *Propellant & Explosives*, Noyes Data Corporation, London, 1974.

30. B. T. Fedoroff, H. A. Aaronson, O. E. Sheffield and E. F. Reese, *Encyclopedia of explosives and related items*, Picatiny arsenal, Dover, N.J., 1974, vol. 1–10.

31. I. Glassman and R. F. Sawyer, *The performance of Chemical Propellants, AGARDograph number 129*, Technivision Services, Slough, England, 1970.

32. R. T. Holzmann, *Chemical Rockets and flame and explosives technology*, Marcel Dekker, New York, London, 1969.

33. F. A. Williams, M. Barrere and N. C. Huang, *Fundamental Aspects of Solid Propellant Rockets*, Technivision Services, Slough, England, Oct. 1969.

34. F. A. Warren, *Rocket Propellants*, Reinhold Publishing Corporation, New York, 1958.

35. *Mechanics and Chemistry of Solid Propellants, Proceedings of the Fourth Symposium on Naval Structural Mechanics held at Purdue University, Lafayette, Indiana, April 19-21, 1965*, Published by Pergamon Press, Oxford, 1967.

36. R. F. Gould, *Advanced in Chemistry Series*, American Chemical Society Publications, 1966.

37. S. F. Sarner, *Propellant Chemistry*, Reinhold Publishing Corporation, New York, 1966.

38. B. Kit and D. S. Evered, *Rocket Propellant Handbook*, Macmillan Company, 1960.

39. R. N. Wimpress, *Internal Ballistics of Solid-fuel rockets*, Mcgrawhill Book Company, USA, 1st edn, 1950.

40. P. M. Deshmukh, D. T. Erande, K. V. Raut, J. Sreenadh, G. T. Kachi and H. Shkehar, Study on Extraction of Ballistic Parameters of 2-Chamber Integrated Rocket with Limited Instrumentation in Static Firing, *Proceeding of 10th International High Energy Materials Conference and Exhibits HEMCE-2016*, 2016, vol. II, pp. 840–843.

41. K. K. Mishra, A. S. Babu and H. Shekhar, Application of Image Processing for Ballsitic Prediction of Star Port Prismatic Rocket Propellants Configurations, *Proceeding of 10th International High Energy Materials Conference and Exhibits HEMCE-2016*, 2016, vol. II, pp. 1079–1084.

42. H. Shekhar, Effects of the Burning Rate Index on the Pressure Time Profile of Progressive Burning Tubular Rocket Propellant Configurations, *Cent. Eur. J. Energ. Mater.*, 2015, **12**(2), 145–158.

43. C. D. Yarrington, S. F. Son, T. J. Foley, S. J. Obrey and A. N. Pacheco, Nano Aluminium Energetics: The Effect of Synthesis Method on Morphology and Combustion Phenomena, *Propellants, Explos., Pyrotech.*, 2011, **36**(6), 551–557.

44. H. Shekhar, Prediction and Comparison of Shelf Life of Solid Rocket Propellants Using Arrhenius and Berthelot Equations, *Propellants, Explos., Pyrotech.*, 2011, **36**(4), P356–P359.

45. K. Jayaraman, S. R. Chakravarthy and R. Sarathi, Accumulation of Nano-Aluminium during Combustion of Composite Solid Propellant Mixtures, *Combust., Explos. Shock Waves*, 2010, **46**(1), 21–29.

46. J. Athar, M. Ghosh, P. S. Dendage, R. S. Damse and A. K. Sikder, Nanocomposites: An Ideal Coating Material to Reduce the Sensitivity of Hydrazinium Nitroformate (HNF), *Propellants, Explos., Pyrotech.*, 2010, **35**(2), 153–158.

47. I. P. S. Kapoor, P. Srivastava and G. Singh, Nanocrystalline Transition Metal Oxides as Catalysts in the Thermal Decomposition of Ammonium Perchlorate, *Propellants, Explos., Pyrotech.*, 2009, **34**(4), 351–356.

48. G. Singh, I. P. S. Kapoor and S. Dubey, Kinetics of Thermal Decomposition of Ammonium Perchlorate with Nanocrystals of Binary Transition Metal Ferrites, *Prog. Astronaut. Aeronaut.*, 2009, **34**(1), 72–77.

49. H. Shekhar and H. Singh, Factors affecting Combustion Behaviour of Metal Powders in Fuel Rich Propellants for Ram rockets: A Fresh Look, *Eighth International Symposium on Special Topics in Chemical Propulsion*, 8-ISICP, Cape Town, South Africa, 2–6 Nov. 2009, pp. 77–79.

50. A. Subhananda Rao and H. Shekhar, Solid Rocket Propulsion Technology in DRDO, *International Workshop on Advances in Processing of Solid Propellant rocket Motors (IWAPS 2008)*, HEMRL, Pune, India, 22–23 Oct. 2008.

51. S. Ozkan, H. Gevgilili, D. M. Kalyan, J. Kowalczyk and M. Mezger, Twin-Screw Extrusion of Nano-Aluminium-Based Simulants of Energetic Formulations involving Gel-Based Binders, *J. Energ. Mater.*, 2007, **25**(3), 173–201.

52. J. Zhi, L. Shu-Fen, Z. Feng-Qi, L. Zi-Ru, Y. Cui-Mei, L. Yang and L. Shang-Wen, Research on the Combustion Properties of Propellants with Low Content of Nano Metal; Powders, *Propellants, Explos., Pyrotech.*, 2006, **31**(2), 139–147.

53. P. R. Patil, V. N. Krishnamurthy and S. S. Joshi, Differential Scanning Calorimetric Study of HTPB based Composite Propellants in Presence of Nano Ferric Oxide, *Propellants, Explos., Pyrotech.*, 2006, **31**(6), 442–446.

54. H. Singh and H. Shekhar, Combustion Mechanism of Solid Rocket Propellants, *5th international seminar on Flame Structure*, Russia, 11–14 July 2005.

55. N. Eisenreich, V. Weiser, H. Fietzek, M. del Mar Juez-Lorenzo, V. Kolarik and A. Koleczko, Mechanicsms of Low Temperature Oxidation for Metal Particle Down to the Nano-Scale and Their Influence to Propellant Combustion, *Proceedings of 31st International Pyrotechnics Seminar*, Fort Collins, Colorado, 11–16 July 2004, pp. 307–317.

56. A. Dokhan, D. T. Bui, E. W. Price, J. M. Seitzman and R. K. Sigman, A detailed Comparison of the Burning Rates and Residual Oxide Products of Ultra-Fine Aluminium in Ammonium Perchlorate Based Solid Propellant, *34th International Annual Conference of ICT*, Federal Republic of Germany, Karlsruhe, 24–27 June 2003, pp. 28-1–28-11.

57. H. Shekhar, *Structural Integrity analysis of case bonded composite propellant*, DRDO Science Spectrum, 2003, pp. P45–P50.

58. P. S. Denade, D. B. Sarwade, S. N. Asthana and H. Singh, Hydrazinium nitro formate (HNF) and HNF based propellants – A review, *J. Energ. Mater.*, 2001, **19**(1), 41–78.

59. V. N. Simonenko and V. E. Zarko, Comparative Studying the Combustion Behaviour of Composite Propelants Containing Ultrafine Aluminium, *30th International annual Conference of ICT*, Federal Republic of Germany, Karlsruhe, 29 June–2 July 1999, pp. 21-1–21-14; M. M. Mench, C. L. Yeh and K. K. Kuo, Propellant Burning Rate Enhancement and Thermal Behaviour of Ultrafine Aluminium Powder (ALEX), *29th International annual Conference of ICT*, Federal Republic of Germany, Karlsruhe, 30 June–3 July 1998, pp. 30-1–30-15.

60. M. J. Chiaverini, K. K. Kuo, A. Peretz and G. C. Harting, Heat Flux and Internal Ballistic Characterization of a Hybrid Rocket Motor Analog, *33rd AIAA/ASME/SAE/ASEE Joint Propulsion Conference and Exhibits*, AIAA, Seattle, WA, July 6–9 1997, pp. 1997–3080.

61. A. Ishihara, M. Q. Brewester, T. A. Sheridan and H. Krier, The Influence of Radiative Heat Feedback on Burning Rate in Aluminized Propellants, *Combust. Flame*, 1991, **84**, 141.

62. H. Singh, Advanced Solid Propellants for propulsion of futuristic Missiles, *1st International Seminar on force Multiplier Technologies for Naval and Land Warfare*, New Delhi, 1999.

63. H. Singh, High Energy Materials Research in India, *J. Propul. Power*, 1995, **11**(4), 848.

64. B. K. Athawale, S. N. Asthana and H. Singh, Metallised Fuel rich Propellants for rocket Ramjet, *Def. Sci. J.*, 1994, **44**, 269–278.

65. S. N. Asthana, C. N. Divekar, R. R. Khare and H. Singh, Evaluation of various dihydric and trihydric phenols as stabilizers for composite modified double base (CMDB) propellants, *J. Hazard. Mater.*, 1991, **27**, 205–211.

66. S. N. Asthana, B. Y. Deshpandde and H. Singh, Evaluation of Various Stabilizers for stability and Increased Life of CMDB Propellants, *Propellants, Explos., Pyrotech.*, 1989, **14**, 170–175.

67. V. K. Bhat, H. Singh and K. R. K. Rao, Processing of High energy Cross-Linked CMDB Propellants, *18th International Conference of ICT*, Karlsruhe, 1987.

68. H. Singh and K. R. K. Rao, Lead Aliphatic Mono and Di-carboxyl-ates as Ballistic Modifiers, *J. Spacecr. Rockets*, 1982, **99**(5), 478–480.
69. K. Kishore and G. Prasad, *Indian J. Technol.*, 1980, **18**.
70. F. Volk, *Proceeding of 5th Symposium on Chemical Problems Connected with Stability of Explosives*, Sweeden, 1979.
71. H. Singh and K. R. K. Rao, *Platonisation in double base rocket propellants, AIAA J.*, 1977, **15**(11), 1545–1549.
72. S. K. Athithan, M. Rajan, M. K. Venkitakrishnan, M. C. Uttam and M. R. Kurup, *Proceeding of Internal Symposium on Space Technology and Science*, Tokyo, 1977.

Subject Index

References to figures are given in *italic* type. References to tables are given in **bold** type.